Second Edition

PRACTICAL HYDRAULICS HANDBOOK

T0136386

Barbara A. Hauser

CRC Press
Taylor & Francis Group
Boca Raton London New York

CRC Press is an imprint of the
Taylor & Francis Group, an **informa** business
A TAYLOR & FRANCIS BOOK

CRC Press
Taylor & Francis Group
6000 Broken Sound Parkway NW, Suite 300
Boca Raton, FL 33487-2742

First issued in paperback 2019

© 1991, 1996 by Taylor & Francis Group, LLC
CRC Press is an imprint of Taylor & Francis Group, an Informa business

No claim to original U.S. Government works

ISBN-13: 978-1-56670-038-2 (hbk)
ISBN-13: 978-0-367-40148-1 (pbk)

Library of Congress Cataloging-in-Publication Data

Catalog record is available from the Library of Congress

Visit the Taylor & Francis Web site at
http://www.taylorandfrancis.com

and the CRC Press Web site at
http://www.crcpress.com

This book is dedicated to those who operate our water utility systems

ABOUT THE AUTHOR

Barbara Hauser has been employed as department head and chief instructor of the Water Purification Technology department at Bay de Noc Community College in Escanaba, Michigan for the past 12 years. Her background includes biochemistry research with the Rockefeller Institute in New York City and at the University of Maine in Orono. She has worked for the Rexnord Company, Milwaukee, Wisconsin as an industrial operations specialist in advanced wastewater treatment, and for the city of Gladstone, Michigan in municipal wastewater treatment. She is a licensed operator in water and wastewater treatment in the states of Michigan and Wisconsin.

In addition to her academic duties, Mrs. Hauser has developed and operated onsite training courses for industrial wastewater treatment personnel, instructed short courses and correspondence courses for waterworks operators sponsored by the Michigan Department of Public Health, organized regional waterworks and wastewater operators training meetings, conducted research for the Michigan Department of Natural Resources, published "Practical Hydraulics Handbook", and "Hydraulics for Operators", and has raised four children.

Mrs. Hauser has an Associate of Applied Science degree from Bay de Noc Community College in Water Technology, a Baccalaureate Degree in Biology/Chemistry from Trinity College in Washington DC, and a Master of Science in Occupational Education from Ferris State University, Big Rapids, Michigan. In 1990, she was awarded the "Outstanding Occupational Educator" award by the Michigan Occupational Dean's Council.

TABLE OF CONTENTS

Page

INTRODUCTION

Controlling the flow of water from one place to another, the proper amount at the proper pressure and velocity, is perhaps more taken for granted than any other aspect of science. Potable water systems must be carefully designed and controlled by the water utility in order to provide adequate pressure, safety from backflow contamination, and proper water velocity for protection of household piping. Water is a carrier for most household and industrial wastes. Proper hydraulic design and maintenance of wastewater collection systems assures sanitary transport to the treatment plant and prevention of sewer backup.

An operator at a water or wastewater utility is most concerned with learning the daily routine of laboratory testing, operation and process maintenance. The hydraulics of the treatment process has been designed into it. It keeps the water moving from one process unit to another, providing correct detention times, proper settling velocity, lift to a higher elevation, etc. Consideration of hydraulic principles may at first be minimal - until the electrical power goes out, a pump malfunctions, a line breaks, or a blockage occurs. When hydraulic control is no longer there, treatment process grinds to a halt. No water utility can operate without it.

In as simple a manner as possible, this text covers the principles and calculations dealing with the hydraulics of water systems. It stresses only what is necessary for a basic understanding, and emphasizes practical applications for water and wastewater utility operations. The text deals essentially with the flow of water, and only occasionally with the control of other liquids or gases.

Calculations are limited to arithmetic conversions, basic algebra, exponential notation. Some previous knowledge in these areas is assumed, as well as knowledge of a few formulas: area of a circle and rectangle, volume of containers of these shapes. A scientific calculator is needed.

Chapters are presented in workbook style, with emphasis on solutions to practical problems. The body of each chapter holds an explanation of the principles to be covered. The applied mathematical section is at the end, and includes a list of problems derived from the concepts. Answers and detailed solutions to each problem are at the back of the text.

I have attempted to set up each solution in as readable a form as possible. In doing so, some shortened forms are in use. The reader should be aware of these:

-- Most values are converted to units of "feet" for calculation. (This includes - feet, square feet, feet per second, cubic feet per second.)

-- Unit labels are often omitted in the body of a solution, but are included with the final answer.

-- Multiplication signs are often omitted in the body of a solution involving a formula.

-- Most answers are rounded off to one or two decimal places.

-- In diagrams, the pressure gage symbol ♀ with a number inside it designates a pressure reading in psi (pounds per square inch).

The second edition of Practical Hydraulics Handbook has added an expanded thrust control section, an updated metering section, and a complete glossary of terms. Many new practice problems are included, and several numeric tables have been developed to ease calculations. Troubleshooting "notes" will enhance calculation accuracy, and the photos are good visual demonstration of hydraulic applications.

CHAPTER 1

MASS, DENSITY AND DISPLACEMENT

mass & weight

Mass is the quantity of matter that a substance contains. It is a basic property of the substance, and is constant, regardless of location. Mass is most frequently registered in units of grams, or pounds; it is measured with an analytical balance which compares it against a known mass.

Weight is the effect of the force of gravity upon a substance, and is measured with a scale. A block of steel which weighs ten pounds on earth, will weigh much less on the moon, where gravitational forces are less. Its mass, however, will be the same in both places.

Since we are dealing only with earthly water systems, for our purposes mass and weight are the same quantity.

density

Also called specific weight, density is mass per unit volume, and may be registered as lb./cu.ft., lb./gal., grams/ml., grams/cu.meter. Taking a fixed volume container, filling it with a fluid, and weighing it, we can determine density of the fluid (after subtracting the weight of the container).

One cubic foot of water weighs 62.4 pounds. (Density = 62.4 lb./cu.ft.)

One gallon of water weighs 8.34 pounds. (Density = 8.34 lb./gal.)

One milliliter of water weighs 1 gram. (Density = 1 gram/ml.)

temperature

The density of materials is affected by temperature. Many solids expand when hot. Fluids also become less dense when warm. Water has the peculiar property of being most dense at four degrees Celsius. Above and below that temperature, it expands. This accounts for seasonal turnover in lakes, and the formation of density gradients, or layers, in reservoirs and behind dams, often concentrating pollution at specific depths. In settling tanks and clarifiers, this effect can inhibit uniform mixing of influent water, causing short circuiting. Most solids will settle more effectively in warm waters than in cold waters because the water is denser when cold.

pressure

Though the density of gases is greatly affected by pressure, the density of most liquids, including water, is not. For all practical purposes, water is considered an incompressible fluid; 62.4 pounds of water occupies a volume of one cubic foot, regardless of the pressure applied to it.

specific gravity

This value designates the ratio of the density of a substance, compared to the density of a standard substance. It is a way to specify relative densities. For liquids and solids, the standard chosen is the weight of water. Therefore:

$$\text{Specific Gravity} = \frac{\text{Density of Substance}}{\text{Density of Water}}$$

If the Specific Gravity of water were to be determined this way, we would write:

$$\text{Specific Gravity} = \frac{\text{Density of Water}}{\text{Density of Water}} = 1$$

The Specific Gravity of water is 1.

Substances with a Specific Gravity of less than 1 will float (gasoline, styrofoam, wood, wastewater scum). Substances with a Specific Gravity of more than 1 will sink (bricks, steel, grit, floc, sludge). A good example of treatment process taking advantage of different Specific Gravities is the multi-media filter. During backwash, the different grades of media are mixed up, but at the end of the backwash cycle, as the water quiets, the media layers become separate and distinct: anthracite on top, sand in the middle, garnet on

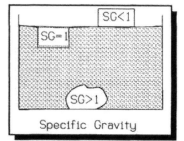

Specific Gravity

the bottom. These substances are chosen for filtration partially because of their different specific gravities.

An anaerobic digester is another good example of the effect of Specific Gravity. Separation of solids, liquid and gas is essential to operation, enabling each to be drawn off separately.

Just as the density of liquids and solids are compared to water as the standard, the Specific Gravity of gases is based on the density of air. Air weighs .075 pounds/cubic foot at standard temperature and pressure. Specific Gravity calculations are done in the same manner, and will determine whether a gas rises or falls in a room full of air.

$$\text{Specific Gravity} \;=\; \frac{\text{Density of Gas}}{\text{Density of Air}}$$

Once Specific Gravity is known, density and weight can be calculated. Note that Specific Gravity is a pure number with no unit label. The units have cancelled out in the calculation.

displacement

A solid submerged in a body of water will displace its own volume. The displacement causes a rise in water level equivalent to the volume of the submerged object. An object partially submerged displaces a volume of water equivalent to the submerged section of the object.

Sometimes it is useful to know the volume of an irregularly shaped object. This can be determined by measuring the displacement. For example, to calculate the percent volume of mudballs in a sand filter, take a core sample with a small cylinder of known volume. Extracting a portion of media, sift out the sand, leaving only the mudballs. Pour them into a large graduated cylinder, filled to a liter with water. A volume of water will be displaced in the cylinder equivalent to the volume of the mudballs. Record the milliliter rise in the cylinder, divide by the volume of the core sampler, and you have the percent mudballs - in the sampler, and in the sand filter.

buoyancy

Buoyancy is known as Archimedes' Principle, and has been employed by man for over two thousand years. An object submerged in a liquid is subject to an upward force equal to the weight of the volume of liquid it has displaced. A dramatic example of this is the underwater swimmer, whose body is lighter than the weight of the water that he is displacing. He actually assumes a negative weight underwater, which causes him to fall upward, unless he purposely swims down. This force of buoyancy applies to any submerged object. For example, a cubic foot of steel weighs 486.7 pounds in air. But immersed in water it weighs 486.7 lb. - minus the weight of the cubic foot of water it has displaced (the force of buoyancy).

PROBLEMS

1.   The water in a tank weighs 820 pounds. How many gallons does it hold?

2.   If the municipal water rate is one dollar per thousand gallons, how many pounds of water are delivered to the customer's house - for a dollar?

3.   Three cubic feet of gasoline weighs 131 lb.
     A. What is the Specific Gravity of gasoline?
     B. How much does a gallon of it weigh?

4.   If an oil weighs 55 lb./cu.ft., how much does a five gallon can of it weigh?

5.   The Specific Gravity of a liquid chemical is 1.27. What is its density (lb./cu.ft.)?

6.   A truck is designed to transport 5000 gallons of liquid.
     A. How many pounds of water can it transport?
     B. How many pounds of sulfuric acid (SG=1.83)?

7.   What is the Specific Gravity of concrete if it weighs 150 lb./cu.ft.?

8.   The Specific Gravity of mercury is 13.6.
     A. What is its weight per cubic foot?
     B. Per cubic yard?

9.   If the density of a sand is 100 lb/cu.ft., how much does a cubic yard of this sand weigh?

10. Chlorine gas is 2.5 times heavier than air.
     A. What is the weight of a cubic foot of chlorine?
     B. What would a room (10' x 12' x 8') full of chlorine gas weigh?

11. The operator is pumping a solution of alum (density = 83 lb./cu.ft.) at a rate of 3 gpm into a chemical treatment system.
     A. How many pounds are fed in one day?
     B. What is the Specific Gravity of the alum solution?

12. A digested sludge whose specific gravity is 1.25 is pumped at 50 gpm to the drying beds.
     A. How many pounds of digested sludge are applied to a bed which is 30 ft. wide, 60 ft. long, and 10 inches deep?
     B. How long does it take to fill this bed?

13. A solid piece of plastic whose specific gravity is 1.2 is dropped from a boat into a lake. How deep will it sink?

14. A gallon of pure mercury is poured into a 24 inch diameter pail. If the mercury weighs 113.4 lb., what is the depth of the mercury in the pail?

15. A wastewater grit chamber measures 5 ft. wide, 30 ft. long, and has a 24 inch water depth. If 22,000 lb. of grit (Specific Gravity = 2.2) were to accumulate in the bottom of this chamber, what would be the new water depth?

16. A stone weighs 90 lb. in air. When immersed in water, it weighs 50 lb.
    A. What is the volume of the stone?
    B. What is its Specific Gravity?

17. If a 150 lb. person swims in a pool, how many cubic feet of water will he displace if his Specific Gravity is 1.1?

18. A heat exchanger is being constructed to maintain temperature of the anaerobic digester. Sludge will pass through a coiled 4 inch diameter pipe 160 ft. long, which is set inside a hot water tank 4 ft. by 4 ft. by 4 ft. high. How many gallons of water will be needed to fill this tank?

19. An open box is to be sunk to its rim in water. If its dimensions are 10 ft. by 10 ft. by 8 ft. deep, how many pounds must it weigh in order to stay submerged?

20. A cubic block of concrete 4 ft. on a side is submerged in a tank of water which measures 6 ft. by 12 ft., and has a water depth of 10 ft. How much does the water level rise?

21. A log with a diameter of 14 inches and a length of 10 ft. weighs 40 lb./cu.ft. If it is inserted vertically into a body of water, what vertical force is required to hold it below the water surface?

22. A liquid with a Specific Gravity of 1.14 is pumped at a rate of 30 gpm. How many pounds per day are being delivered by the pump?

23. A core sample of the sand filter is taken to check for percent mudball accumulation. The sampler is 6 inches in diameter and 12 inches long. After sifting out the sand, the mudballs are immersed into a 2 inch diameter graduate holding 1 liter of water. The water level in the graduate rises 4 inches. What is the percent mudballs in the filter?

24. A long narrow cylinder weighing 25 pounds was filled with water and placed on a scale. A 2 inch diameter probe hanging from the ceiling was submerged 3 feet in this water. If the scale registered 300 pounds, how many gallons of water were in the cylinder?

25. A cubical float, 4 ft. on a side, weighs 400 lb. and is anchored by means of a concrete block which weighs 1500 lb. in air. Nine inches of the float are submerged when the chain connected to the concrete is taut. What rise in the water level will lift the concrete off the bottom? (concrete density = 150 lb./cu.ft.)

CHAPTER 2

FLOW AND VELOCITY

Flow - the quantity of water passing a point in a given unit of time. Think of flow as a volume - which is moving. It can be recorded as gallons/day (gpd), gallons/minute (gpm), or cubic feet/second (cfs). Flows treated by a municipal utility are large, and often referred to in Million Gallons per Day (MGD).

Flow may be enclosed in pipes, or it may be open channel flow. The first 8 chapters of this text will be considering predominantly closed pipe flow. Pipes are flowing full, and under pressure. Open channel flow deals with conduits which are partially full, and the water surface is in contact with atmospheric pressure (streams, aqueducts, sewer pipes, grit chambers, clarifiers). The water flows because of the slope of the conduit. Open channel flow will be taken up later in the text.

Flow may be laminar or turbulent. Laminar flow only occurs at extremely low velocities. The water moves in straight parallel lines, called laminae, or streamlines, which slide upon each other as they travel, rather than mixing up. Turbulent flow, which is normal pipe flow, occurs because of friction encountered on the inside of the pipe. This throws the outside layers of water into the inner layers; the result is that all the layers mix and are moving in different directions, and at different velocities. Added up, however, the direction of flow is forward.

Flow may be steady or unsteady. We will be considering steady state flow only. At any one point, the flow, and velocity, does not change. Most hydraulic calculations are done under the assumption of steady state flow. If we can assume a given flow at a given point, calculations can be based on that fact. If we had to consider that the flow was constantly changing, calculations would become extremely complex. Note, however, that in actual systems, flow is frequently in an unsteady condition. As a water tank empties into a pipe, the flow through the pipe decreases over time, because the depth of water in the tank which pushes the water forward is decreasing. As a pump fills a tank from a bottom entrance, that pump is delivering less water as the tank gets fuller and fuller. Flow through a wastewater treatment plant is constantly fluctuating, but process treatment units are designed for an average, maximum or minimum flow, depending on the need. It is understood that unsteady flow is a common condition, but this is not normally dealt with in calculations.

equation of continuity

Under the assumption of steady state flow, <u>the flow that enters the pipe is the same flow that exits the pipe</u>. It is continuous. Water is incompressible; it cannot accumulate inside. The flow at any given point is the same flow at any other given point on the pipeline. This is true in any water system, as long as no additional flows are added, and no exits which split or divert the flow away in other directions are present.

However, the velocity of the water may change. At a given flow, <u>the velocity is dependent upon the cross sectional area of the conduit</u>. Velocity is the speed at which the flow is traveling. A given quantity of water will travel faster through smaller spaces, and slower through larger spaces. Consider this phenomenon in a closed pipe. A 50 gal./min. flow travels at a very slow velocity through a 36 inch diameter pipe, but if the pipe narrows to 4 inches diameter, that same 50 gal./min. will be moving very fast through the narrow section, in order for the flow to remain the same.

The principle applies equally to open channel flow. Water entering a wastewater treatment plant travels at scouring velocity coming down the pipe (2-3 ft./sec.). Entering the grit chamber, which is larger in cross sectional area than the sewer pipe, the velocity slows down to about 1 ft./sec. to allow grit to settle. The next process unit, the primary clarifier, again takes advantage of this principle, and is

Figure 2.1 - Pipe reducer produced velocity change.

large enough to slow the velocity to about .05 ft./sec. so that organics will settle. Much of wastewater treatment is designed upon this principle, and detention times are developed from it.

The <u>Equation of Continuity</u>:

$$Q = A \, V$$

Q = Flow
A = Area (cross sectional area of conduit)
V = Velocity

This is the most basic hydraulic equation. It always holds true, and all other hydraulic formulas determine components of this one.

Be aware that if connected pipes of different diameters are considered, substitution within the formula can be performed to shorten calculations.

If: $Q_{pipe\ 1} = Q_{pipe\ 2}$

Then: $A_1 V_1 = A_2 V_2$

Note: The unit labels used must be in the same dimensions across the formula. It is easiest to label flow as cu.ft./sec. (cfs), area as sq.ft., and velocity as ft./sec. If values are presented in other dimensions, first change them to these before inserting into the formula.

PROBLEMS

1. A flow of 1 MGD occurs in an 8 inch diameter pipeline. What is the water velocity?

2. What is the velocity of water in a 36 inch diameter pipe which is carrying 50 gpm?

3. A city requires a flow of 40 MGD. What diameter pipe is required to carry this flow, if the water velocity is to be 4 ft./sec.?

4. The cross sectional measurements of a flume are 10 inches by 10 inches. The water is 8 inches deep, and moving at a velocity of 2 ft./sec. How many gallons of water will the flume deliver in four hours?

5. A 22 inch diameter pipe carries water at a velocity of 150 ft./min. What is the flow rate (gpm)?

6. Wastewater flows into a sewer trunk through a wye at 350 gpm. If the trunk flow was 2.2 cfs upstream of the wye, what is the trunk flow downstream of the wye (MGD)?

7. What capacity pump would be necessary in a sewage lift station to maintain a velocity of 3 ft./sec. in the 4 inch force main discharge pipe?

8. The velocity of flow in a 12 inch diameter pipe is 100 ft./min. The flow discharges through a nozzle 3 inches in diameter. What is the velocity of flow in the nozzle?

9. A rectangular open channel is 5 ft. wide and carries a flow of 12 MGD. A small child places his sailboat in the channel, and it travels a mile in 30 minutes. How deep is the channel?

10. How many gallons of water can be stored in a pipeline 5 ft. in diameter and 4 miles long?

11. A channel 4 ft. wide has water flowing to a depth of 2 ft. What is the gpm flow in the channel if the water travels at a speed of 3 ft./sec.?

12. At a manhole, dye is injected into a 36 inch diameter sewer which is flowing half full. The dye appears at the next manhole, 350 ft. downstream in 2.5 minutes. What quantity of water is passing through this pipe (gpd)?

13. A rectangular channel 24 inches wide and 18 inches deep is flowing half full. What is the discharge (cfs) when the velocity is 12 ft./min.?

14. An 8 inch diameter pipeline carries water to an industry at a velocity of 10 ft./sec.
    A. What is the flow in gpd?
    B. If water is purchased at $1.00/1000 gal., what is the quarterly bill for water usage?

15. Water exiting a tank enters a 24 inch diameter pipe at 5.6 cfs. At 250 feet of length the pipe narrows to 22 inch diameter and the velocity increases by .35 ft./sec. at this section. What is the cfs flow at the downstream end of the 22 inch diameter section?

16. Water travels through a 45 ft. long grit chamber at a velocity of .92 ft./sec. If the grit chamber length is extended to 60 ft., what will be the water velocity?

17. A 24 inch diameter pipe with a water velocity of 60 ft./min. discharges into a rectangular clarifier 10 ft. deep, 10 ft. wide and 100 ft. long. What is the water velocity in the clarifier?

18. Flow = 10 MGD
    Find velocities A, B, C

19. Presented is part of a pipe system.
    A. What is the diameter of pipe A?
    B. What is the velocity in pipe C?
    C. What is the flow in pipe D?

20. A 100 gpm pump:
    A. Will pump how many cfs?
    B. Will pump how many cu. meters/day?

21. A rectangular channel constricts from a width of 3 ft. to a width of 2.5 ft. in a short transition section. If the flow is 6.5 cfs and the depth upstream of the constriction is .4 ft., what is the depth downstream of the constriction.

22. A pipe 12 inches in diameter reduces to a diameter of 6 inches, then expands to a diameter of 10 inches. If the average velocity in the 6 inch diameter pipe is 15 ft./sec., what is the average velocity at the other sections?

23. A chlorine contact tank has a capacity of 6000 gal. and a length of 20 feet. Calculate the velocity in the tank when the flow is .432 MGD.

24. A 6 inch diameter pipe carries 2.87 cfs. The pipe branches into a 2 inch diameter pipe and a 4 inch diameter pipe. If the velocity in the 2 inch diameter pipe is 40 ft./sec., what is the velocity in the 4 inch diameter pipe?

25. Water flows through a trickling filter at a rate of 2 MGD. A portion of the water leaving the filter is being recirculated back to the filter, flowing through an 8 inch diameter pipe at a velocity of 2 ft./sec. What is the flow entering the headworks of this plant?

CHAPTER 3

PRESSURE

As all fluids, water has a specific weight: 62.4 pounds for every cubic foot, or 8.34 pounds for every gallon. This weight, resting on a surface, exerts a force on that surface. For example, one cubic foot of water, resting on its bottom surface, exerts a force of 62.4 pounds on that square foot (62.4 lb./sq.ft.). Force on a unit area is called Pressure.

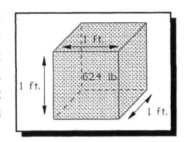

If the weight were composed of two cubic foot blocks of water, one on top of each other, then the specific weight would be 62.4 lb./cu.ft., the height would be 2 feet, and the pressure on the bottom surface would be 124.8 lb./sq.ft. Density remains constant, but the pressure on the unit area depends on the height, or depth of the water.

A formula develops:
Pressure = Weight x Height
124.8 lb/sq.ft. = 62.4 lb/cu.ft. x 2 ft.

If there were four blocks of water, arranged as in the diagram to the right, the pressure on every square foot of bottom surface would still be 124.8 lb./sq.ft. It would not matter how wide the water was. It is just the height, or depth, that is pushing down on that surface, and causing the pressure.

It is most convenient to specify pressure at a particular point in a water system. For example, what is the water pressure at a specified point along a pipeline? Or, what is the pressure at a particular point on the bottom of a swimming pool? Using the concept of pressure, there is a need to deal with a very small unit area. Therefore, pounds per square inch (psi) has become the favored designation.

> **Note:** The unit label for specific weight (density), is a weight per volume designation (lb./cu.ft.).
>
> The unit label for Pressure is a weight per area designation (lb./sq.ft.). It applies to the surface area that the weight is resting on.

Taking another look at the cubic foot block of water, in order to register pressure as lb./sq.in., the bottom surface can be split into 144 square inches. Now, how much water is resting on each of those square inches? Picture a column of water rising from each of them to the top of the block. How much does each column weigh?

$$\frac{62.4 \text{ lb./cu.ft.}}{144 \text{ sq.in./sq.ft.}} = .433 \text{ lb. per foot of height}$$

<u>A cubic foot block of water, one foot high, exerts a pressure of .433 lb. on every square inch of bottom surface area.</u>

The double block of water, two feet high, exerts a pressure of .866 lb. on every square inch of bottom surface area. In all cases, the pressure depends upon the depth of the water above it, and an equivalence of feet of depth to psi pressure is created.

However, there is a more comfortable way to state this equivalence. For practical purposes, the water utility superintendent would like to know how many feet of water depth must be provided in the city water storage tank in order to have adequate pressure in the pipes. There is needed an equivalence of feet of water depth to psi pressure that is readily interchangeable, and which will generate discrete units of psi pressure.

Creating a ratio, if a 1 ft. column of water is equivalent to a pressure of .433 psi, then how many feet of water will equal 1 psi?

$$\frac{.433 \text{ psi}}{1 \text{ ft.}} = \frac{1 \text{ psi}}{x \text{ ft.}}$$

$$x = 2.31 \text{ ft.}$$

Therefore, we see that <u>1 psi is equivalent to a 2.31 ft. depth of water</u>. Water pressure can be registered as psi, or as feet of water. The conversion 1 psi = 2.31 ft. is most commonly used. However, .433 psi = 1 ft. will yield the same result in calculation.

pascal's law

In the 17th century, Blaise Pascal formulated a fundamental law of hydraulics.

<u>The pressure at any one point in a static liquid is
exerted with equal intensity in all directions</u>.

For instance, the pressure at any point on the horizontal bottom of a tank of water is the same, and the pressure exerted upward, downward, and to the sides from any one point is the same also. Dewatered basement floors in flooded areas buckle from this upward pressure, and structures built below ground, such as dewatered tanks at water or wastewater treatment plants, must be able to withstand upward pressures imposed by groundwater, or they may float right up out of the ground.

height vs. shape

Pressure in a static system is due to the depth of water above the point of measurement.

Pressure is exerted with equal intensity in all directions.

Keep both of these facts in mind when inspecting the diagram below. The source of distribution system pressure in most community water systems is the water storage tank.

The tanks may take a number of shapes, but these same basic hydraulic principles apply. In each of the diagrams to the right, as long as the depth of the water is the same from the water surface to the point of measurement, the water pressure at that point is the same. The shape of the tank does not matter, nor does the number of gallons each holds.

#1 is a typical elevated tank. Let us say that it holds 400,000 gallons of water when the water level is 100 feet above ground surface. A pressure gage at ground level will register 43.3 psi, because of the 100 ft. depth of water above it. The pressure registers downward from the water surface, straight through the riser pipe to the gage.

#2 is a standpipe, also containing 400,000 gal. when the water surface is 100 ft. above ground. It will also register 43.3 psi at the bottom. The difference between these two types is in the use of the water. The water pressure which supplies the community from #1 will remain high until the tank is empty. #2 provides the same pressure when full, but as water is used and the level drops, the pressure will drop significantly with it.

#3 is a ground level storage tank, set into a hill, with a pressure gage on the pipeline 100 ft. below the water surface. The pressure gage will register 43.3 psi. The community is down at this level, and the water storage tank was installed at a higher elevation to provide the pressure. It doesn't matter that the tank is buried in the ground instead of being up in the air. The same depth of water is still provided by the direct physical connection of water from the tank through the pipeline. The pressure registers down, and to the side, all the way down the pipe to the gage location.

#4 is a copy of #2, equipped with a ½ inch diameter sight glass, to read the water depth. Connected to the bottom of the tank, the water will rise in the glass to the water level in the tank - 100 ft. The gage at the bottom of the tank registers 43.3 psi. If a valve between the sight glass and the tank is then closed, a gage at the bottom of the sight glass will still register 43.3 psi. Though there is only a little water in that glass, it is 100 ft. deep above the gage.

Figure 3.1 - Elevated Water Storage

pumps

Adequate elevation and depth of water in a tank will provide enough pressure to move the water throughout the distribution system. When elevation is not available, a pump is used to provide the needed pressure. The pump is actually providing lift (elevation), or enough pressure to raise the water to a given height. If the direction of flow is horizontal from the pump discharge, then the pressure remains in the discharge pipes, and keeps the water moving. Pumped pressure can also be registered in feet of water, or in psi.

pressure measurement

Piezometer

The simplest type of pressure measurement device is the piezometer. This is simply a narrow, open-ended vertical tube, attached at the point of pressure measurement. Water will rise in the tube to a height equivalent to the pressure. If a piezometer is attached to the bottom of a tank, and is made of glass, it is referred to as a sight glass, and provides an easy means of reading the water level. The water depth can easily be converted to psi to obtain a pressure reading at the bottom. This is practical for small tanks of water at ground level, but obviously impractical on an elevated storage tank. Piezometers can be installed on pipelines; the water will rise in the tube to a level equivalent to the pressure in the pipe. As long as the piezometer is tall enough, the water will not pour out the top. It will rise till the depth of water in the tube provides a downward pressure equivalent to the upward pressure in the pipe. Again, this is most often impractical. A pipeline under a pressure of 60 psi, would need a piezometer 138.6 ft. high.

Pressure Gage

The pressure gage is a compact, practical device used for pressure measurement. It is probably the most important instrument in the water system. Called a <u>Bourdon Tube</u>, it should be calibrated in feet of water, though it may read feet, psi, or inches of mercury. The dial size is usually 4½ inches in diameter, for readability, and the range selected should go beyond the maximum operating pressure of the system. The needle can be zeroed with a screwdriver. The gage assembly is a liquid filled system. The snubber is a restrictive device that stops movement of the liquid fill, deadening pulses of power which cause a blurred pointer motion. Glycerine is often preferred for fill, and prevents solids bearing water from damaging the gage. The diaphragm seal isolates the water from the fill, but allows the pressure to be transmitted through. The bleed screw on the bottom of the diaphragm seal is for bleeding air if the pointer does not return to zero when the gage

is disconnected. A ball valve should be at the bottom for shutoff. The nipple is the connection of the gage to the pipe, and should be welded on, or attached with a service saddle (not tapped in). When this area becomes clogged, it is easy to remove the gage above the valve, and rod out the nipple and valve.

gage position

A pressure gage is usually attached directly onto the pipe, and reads pressure directly from it. Occasionally, for ease in reading, the gage is placed at some distance from the pipe, either above or below it. If an automatic correction for the difference in elevation is not built into the apparatus, it is easy to calculate. A gage placed below the pipeline will register a higher pressure than the pressure in the pipe. The water extends down to the diaphragm seal on the gage. If the extension is 5 ft. long, then the gage will register the pipe pressure, plus the extra five feet of elevation. If the gage is above the pipeline, it will register the pipe pressure minus the distance between the gage and the pipe. In this case the pipe pressure is more than the gage pressure.

> **Note:** Change pressure units from psi to feet when calculating in these distances.

Keep in mind that even if the water in the pipe is flowing, the water exiting to the pressure gage is static. It dead ends at the gage; this water is not moving, and except for the difference in elevation from the pipe to gage, the actual pipe pressure at the point of connection is being transmitted to the gage. Bends and turns in the gage extension line will not affect the gage reading (Pascal's law-pressure registers in all directions). Only the elevation difference between gage and pipeline is significant.

air pressure/water pressure

The pressure read on a pressure gage is referred to as Gage Pressure. It is the difference between a given pressure and that of the atmosphere. It is most often a positive number, but can also be negative, such as in siphon lines, or on the suction side of a pump which draws water up from below. Some gages are calibrated to read negative pressures, and register it with a minus sign. These gages are reading the amount of pressure below atmospheric pressure. They are actually reading the amount of vacuum in the line.

Atmospheric Pressure

The atmosphere is composed of gases. They have weight, as liquids do, and exert a pressure upon the earth. We don't feel this pressure, because we are so used to walking around in it, but the atmosphere, about 200 miles deep, exerts a pressure of approximately 14.7 psi at sea level, the standard reference point for all pressure measurements. It exerts that same pressure on an open tank of water. In fact, it is this pressure that keeps water in a liquid state. If a tank of water were sealed, and all the air pressure vacuumed out of it, the water would

become vapor. It would no longer have the force upon it which holds it together as a liquid. We experience some evidence of this on days when a zone of low atmospheric pressure enters our area; water evaporates more easily, creating cloudy, humid days. Weather stations record atmospheric pressure on a barometer, a piezometer filled with mercury (atmospheric pressure = 30 inches mercury). If we could vacuum all the air out of a container, the pressure reading on an attached gage would be -14.7 psi, or -34 feet of water: Absolute Vacuum. This is equivalent to an Absolute Pressure reading of 0.

Absolute Pressure

Pressure readings can be reported either as Absolute Pressure or as Gage Pressure. The difference is that in Gage Pressure readings, atmospheric pressure is not counted. With Absolute Pressure readings, it is.

Absolute Pressure (psia) = Gage Pressure (psig) + Atmospheric Pressure (14.7 psi)

Gage Pressure is the pressure above or below atmospheric, and is recorded as psig. It can be negative - down to minus 14.7. The Absolute System of measuring pressures is less commonly used in hydraulic calculations and is mostly reserved for pumping applications; this text will deal only with Gage Pressure until otherwise indicated.

Note: Absolute Pressure is always positive. It is theoretically impossible to have an Absolute Pressure reading below zero.

---

total pressure - force

Force is the pressure on an entire area. The force downward on the entire bottom of a tank filled with water, laterally on the side of a dam, or upward on the cover of an anaerobic digester. Force is registered in pounds, and is calculated by multiplying the pressure (lb./sq.ft.) times the area (sq.ft.). The formula:

Force = Pressure x Area

lb. = lb/sq.ft. x sq.ft.

can be used with other unit labels (square inches) as well, as long as the same dimensional units are carried across the entire formula.

One cubic foot of water exerts a <u>pressure</u> of .433 lb. on each square inch of bottom surface. It exerts a <u>force</u> of 62.4 lb. on the entire square foot of bottom surface.

When calculating force on a vertical surface, such as the side of a tank, an average must be taken. The pressure at the water surface is zero; the pressure at the bottom is maximum; therefore, an average of the two must be used, and then multiplied by the sidewall area.

Force also becomes a significant factor in treatment processes when dealing with process tanks which are set below ground surface. If the water table is high in the area, it may be necessary to dewater the area around the tank before dewatering the tank for cleanout or repairs, or the surrounding force of water exerted upward on the bottom of the empty tank will pop it right out of the ground.

When a force is applied to a gas, the gas is able to absorb the pressure, and compresses to a smaller volume. When the pressure is released, more space is available, and the gas expands to fill the entire space. For practical purposes, liquids do not have this capacity. Water is an incompressible fluid. When pressure is applied, it will not condense; volume remains the same, and pressure is transmitted to the surroundings. This yields dramatic results when moving water comes to a sudden halt.

effects of pressure - dynamic systems

Water Hammer

Called Hydraulic Shock, water hammer is the momentary increase in pressure which occurs in a moving water system when there is a sudden change of direction or velocity of the water.

When water flowing in a pipeline is suddenly stopped by a rapidly closing valve, pressure energy is transferred to the valve and pipe wall. Shock waves are set up within the system. Waves of pressure travel backwards until encountering the next solid obstacle, then forward, then back again. Neither the pipe nor the water will compress to absorb the shock. Water is 100 times less compressible than steel; this sudden increase in pressure will damage pipes, valves, and shake fittings loose. The velocity of a pressure wave is equal to the speed of sound; therefore it "bangs" as it travels back and forth, until dissipated by friction losses. We are familiar with the "bang" that resounds throughout an older house when a faucet is suddenly closed - an effect of water hammer.

Air chambers are installed in areas where water hammer is encountered frequently, and are typically seen behind sink and tub fixtures and on sludge handling piston pumps. Shaped like thin upside-down bottles, with a small orifice connection to the pipe, they are air filled. The air compresses to absorb the shock, protecting the fixture and piping. On centrifugal pumps, which can be damaged by hammer caused when electrical power fails, the best form of prevention is to have automatically controlled valves which close slowly.

A less severe form of hammer is called Surge, a slow motion mass oscillation of water caused by internal pressure fluctuations in the system. This could be pictured as a slower "wave" of pressure building within the system. In long pipelines, surge can be relieved with a surge tank, a tank of water directly connected to the pipeline. When surge is encountered, it will act to relieve the pressure, and can store excess liquid. Surge tanks can serve for both positive and negative pressure fluctuations.

Both water hammer and surge are referred to as transient pressures. They both yield the same results if not controlled-damage to pipes, fittings, valves, causing leaks and shortening the life of the system.

Excess pressure in water lines can also be caused by entrained air, or by temperature changes of the water. Air trapped in the line will compress, and will exert extra pressure on the water. Temperature changes will actually cause the water to expand or contract, also affecting pressure. These conditions can be controlled by pressure relief valves, set to open with excess pressure in the line, and then close when pressure drops. Usually spring loaded, they are referred to as "surge suppressors". These are an integral part of a heating system, along with air chambers to hold excess water at high temperatures.

Figure 3.2 - Surge chambers installed on pump to prevent hammer.

Thrust

Thrust is the force which water exerts on a pipeline as it turns a corner. The phenomenon is similar to water hammer, but it is constant force. Caused by an imbalance of internal pressures, the intensity of thrust is dependent upon water momentum, and is greater at higher velocities and in larger pipelines. Thrust forces act perpendicular to the outside corner of the pipe, and affect bends, tees, reducers, dead ends, pushing the coupling away from both sections of pipeline. Uncontrolled, thrust will separate the coupling and cause leakage.

To calculate Total Pounds of Thrust:

$$\text{Thrust} = 2TA \times \sin\tfrac{1}{2}\,\theta$$

Where: T = test pressure of system in psf, plus 100 psi (14400 psf) for water hammer. Test is usually 150 psi (21600 psf).
A = cross sectional area of fitting (sq.ft.)
$\sin\tfrac{1}{2}\,\theta$ = sine of half the angle of turn
(most of the thrust occurs half way into the turn)

For pipes using push-on or mechanical joints, thrust restraint is desired, since neither of these joints provide significant restraint against pipe separation except for the friction between the gasket and the plain end of the pipe. The design objective of the restraint device is to distribute thrust forces to the soil structure to avoid separation and damage to the pipe system.

Thrust Blocks - One of the most common methods of providing resistance to thrust forces is the use of thrust blocks. These are massive concrete blocks with a crushing strength of at least 2000 psi which are cast in place onto the pipe and around the outside corner of the turn. Thrust is transferred to the soil through the larger bearing area of the block. Blocks should be placed in undisturbed or compacted soil, block width is often twice the

height of the block, the section cast around the pipe should not overlay the bolts or the joint connection, and the bearing

Soil Bearing Strengths	
soft clay	1000 lb/sq.ft.
silt	1500 lb/sq.ft.
sand	4000 lb/sq.ft.
clay	9000 lb/sq.ft.

pressure of the block against the soil must not exceed the bearing strength of the soil (or the soil will move). Soil bearing strengths vary.

To calculate Bearing Face Area of the thrust block:

$$\text{Area (sq.ft.)} = \frac{\text{Total Thrust (lb)}}{\text{Bearing Strength of soil (lb/sq.ft.)}}$$

Gravity thrust blocks (thrust anchors) are used for downward bends; they are shackled to the fitting with tie rods and are made of specific density material to make full use of gravity to hold the fitting in place.

<u>Hydrants</u> are often placed on a cornered set of concrete pads, one beneath for leveling and one in back for anchorage. Hydrant fittings may have special restrained joints.

Restrained Joints - In locations where it is difficult to use thrust blocks (if the soil has poor hearing strength, if there is no space for a thrust block, or if thrust forces are not expected to be great and economy is desired), restrained joint pipe is an alternative. The restrained joint transfers the thrust load from the pipe directly to the surrounding soil. At times it is necessary to restrain the joints also along the lengths of pipe on both sides of the location of thrust. The amount of thrust ex⁻ ːcted and the bearing strength of the surrounding soil will determine the number of joints and the length of pipe which needs restraining.

Figure 3.3 - Thrust block in above ground application.

For push-on joints, the bell is manufactured with an extension which locks onto a snap ring inserted on the spigot. The two pipe segments are actually locked together.

Restrained Joint

For mechanical joints, a retainer gland is inserted on the spigot and bolted to the mechanical joint bell. Extra long bolts are used; they hold the connection more stably under thrust than the mechanical joint alone, yet allow considerable joint deflection.

Insertion of concrete curbing along with restrained joints is sometimes practiced, to add extra stability. At the connection of a water main to a lateral, a tapping sleeve may be installed to keep the joint stable.

Tie Rods - Tie rods can be used to restrict pipe movement in above ground installations. For mechanical joints, tie rods can be threaded through the bolt holes in the flange; is some cases, more than one length of pipe on each side of the fitting may require restraint.

Tie rods

PROBLEMS

1. A pressure gage at bottom of a standpipe reads 105 psi. What is the depth of water in the standpipe?

2. House elevation = 2150 ft.
   Gage A elevation = 2222 ft.
   Water surface elevation = 2290 ft.
   A. With valve A open and valve B closed, what does gage B read (psi)?
   B. With valve B closed and valve A closed, what is the reading at B?

3. A 100 ft. diameter cylindrical tank contains 1.5 MG water.
   A. What is the water depth?
   B. What pressure does a gage at the bottom read (psi)?

4. The pressure in a pipe is 65 psi.
   A. What is the pressure in feet of water?
   B. What is the pressure in psf?

5. What is the static pressure (psi) two miles beneath the ocean surface?

6. A half full standpipe on the bluff (elevation 2248) is 150 ft. tall. At a customer's residence downstream from the standpipe the water pressure registers 62 psi on a gage. What is the elevation of the customer's house?

7. What height of a column of oil (SG=.84) would exert the same pressure as a column of water 46 ft. high?

8. What is the pressure of a column of mercury (SG=13.6) which is 10 ft. high, on one square inch of its bottom surface?

9. The pressure in a pipeline is 6234 psf. What is the head on the pipe?

10. A gage on the suction side of a pump shows a vacuum of 10 inches of mercury.
    A. What is this pressure in feet of water?
    B. What is the pressure in psi?

11. A pressure gage positioned 7.5 ft. above a pipeline reads 200 psi. What is the pressure in the pipe?

12. Existing 30 psi pressure in a pipeline is transmitted through 300 ft. of tubing to a gage which is 10 ft. below the pipe. What is the pressure reading on the gage?

13. What must the pressure in a water pipe 30 ft. below a closed faucet be to give a pressure of 35 psi at the faucet?

14. The pressure on a surface is 37 psig. If the surface area is 1.8 sq.ft. what is the force (lb) exerted on the surface?

15. A barge 300 ft. wide, 600 ft. long loaded with iron ore pulls a draft of 25 ft. What is the upward force (lb.) of the water on the bottom of the barge?

16. A common vertical wall between two tanks of liquid measures 20 ft. high and 20 ft. wide. Water in one tank is 15 ft. deep. Oil (SG=.77) in the other tank is 6 ft. deep. What is the force acting on the wall?

17. A triangular trough carrying water is 200 ft. long. What is the total stress on the side of the trough if water in it is 3 ft. deep, and the width of the water surface is 8 ft.?

18. A concrete gravity dam with a vertical upstream face is 80 ft. high. What is the total force on the dam when the water level is 4 ft. below the top of the dam? Dam width at river bed is 30 ft. and sides slope at 45 degree angle outward vertically.

19. A water storage tank on top of a building supplies the cylindrical hot water heater in the basement, 50 ft. below. The water storage tank has 15 ft. of water in it. The hot water heater is 6 ft. tall and has a diameter of 4 ft. What is the average force on the heater walls?

20. A tank full of water measures 15 ft. long, 10 ft. wide and 12 ft. high.
    A. What will a pressure gage at the bottom read (psi)?
    B. What is the force of the water on the bottom?

21. A tank is 10 ft. by 15 ft. by 25 ft. water depth. What is the pressure at the bottom:
    A. in psig
    B. in psia
    C. in psfg
    D. in psfa

22. A thrust block is being cast onto a 12 inch diameter pipe at a 45 degree bend in sandy soil. What should the area of the bearing face of the block be?

23. A closed rectangular steel box 40 ft. by 20 ft., and 20 ft. high, and weighing 80 tons, is to be sunk in water.
    A. How deep will it sink when launched?
    B. If the water is 18 ft. deep, what weight must be added to sink it to the bottom?

24. An iceberg has a specific weight of 57.2 lb./cu.ft. What portion of its total volume will extend above the surface if it is in fresh water?

25. A. A basement which measures 35 ft. by 60 ft. by 7 ft. high has its floor 6 ft. below grade. Assuming the total weight of the house and basement is 1,250,000 lb., what elevation of the groundwater would be critical for this house?

   B. The owner now paints the house with a "waterproof" paint which successfully resists a pressure equivalent to a 96 inch column of oil (SG=.80). Will the basement leak before it floats? Why?

CHAPTER 4

BERNOULLI'S THEOREM

static systems

From the preceding chapter, we have seen that when a vertical tube, open at the top, is installed onto a tank of water, the water will rise in the tube to the water level in the tank. This tube is called a piezometer. The water level to which the water rises in a piezometer is the piezometric surface.

Now consider a pipeline under a constant source of pressure, with a closed valve at the end (static system). If several piezometers are installed onto the pipe in series, the water will rise in each of the piezometers to a height which is equivalent to the pressure in the pipe, and it will be at the same level in each of them; the system is static. If the pipeline pressure is 40 psi, the water will rise to 40 psi x 2.31 ft./psi: 92.4 ft. in each piezometer. It will not pour out the top. The pressure in the pipe is equivalent to 92.4

ft. of head. Once the water in each piezometer reaches 92.4 ft., its weight pushing down upon the water in the pipe is equal to the pipe's pressurized water pushing upward, and it stops rising. The width of a piezometer is of no significance. The water will rise to the same level no matter how wide it is.

The Piezometric surface registers the level of pressure energy in the pipeline, and this level is designated all along the pipe length by an imaginary line called the Hydraulic Grade Line.

The Hydraulic Grade Line is the line that connects the piezometric surfaces along a pipeline.

In the diagram to the right the pressure source is included, the full tank of water, with

its own piezometer attached. The pressure in our pipeline originated directly from the depth of water in this tank, and each of the piezometers along the pipeline is registering this pressure. Note the closed valve on the pipe end, designating a static water system.

Note: In a static water system, the Hydraulic Grade Line is always horizontal.

The pressure does not change along the length of that pipeline, because the water is not

moving and there are no pressure losses. If the pipe itself is not horizontal, but curves up and down, the Hydraulic Grade Line will still remain horizontal, as long as the system is static. The pipeline pressure will be lower where the pipe is elevated, and higher where the pipe is at a lower level.

The concept of a piezometric surface is not only demonstrated in man-made pipe systems, but also in nature, with groundwater. A well, which is actually a piezometer, is drilled into the aquifer. If this is a water table aquifer, and not under pressure, water will enter the well to the level of the water table. If the well has been drilled into an artesian aquifer, under pressure, then the water will rise in the well to a distance equivalent to the pressure in the aquifer at that point, possibly all the way to the surface.

dynamic systems - head loss

In a Dynamic Water System, where water is flowing, the piezometric surface varies. The Hydraulic Grade Line slopes downward along the length of pipe, showing a loss of

pressure energy. The difference between the level of the Hydraulic Grade Line in the static system, and that in the dynamic system, at any one point, is the feet of pressure lost because of friction in the pipe. This is referred to as <u>head loss, and is an important factor</u>. A water utility must account for these pressure losses in order to provide adequate pressure service throughout the system, when determining the water level in the municipal water tower.

Obviously, it is not practical to install piezometers on water towers, or on pipelines. In some cases, they would have to be hundreds of feet high. Instead we use pressure gages, which record pressure in feet of water, or in psi.

types of head

We know that pressure at a given point originates from the height, or depth of water above it. It is this pressure, or head, which gives the water energy, and causes it to flow. The components of pressure are:

Pressure Head

This is the pressure which is directly due to the depth of the water. If a tank of water is 80 ft. deep, the head is 80 ft. at the bottom of the tank and the psi pressure is 80 ft./2.31 ft./psi = 34.6 psi. The same pressure registers at all points along a horizontal connected pipeline, as long as the system is static. If it is a dynamic system and water is flowing, the pressure will be 34.6 psi at the bottom of the tank, and will decrease over the length of the pipeline due to friction losses in the pipe.

Note: Pressure Head, when used in calculations, is always registered as feet of water.

Elevation Head

If the tank of water (80 ft. deep) were up on a hill, higher in elevation than the point of measurement, the pressure in the static system would still be measured from the water surface. (HGL is always horizontal in a static system). From the point of measurement at the lower elevation, a pressure gage reading would include the depth of the water, and the elevation of the tank. This pressure due to elevation, registers on the gage and is called Elevation Head.

Water storage tanks are often placed at high elevations in order to take advantage of Elevation Head. Some large city water utilities have chosen to draw their raw water from more distant locations, rather than a closer source, because of the available Elevation Head to provide added pressure. Most cities have elevated water storage towers, which also employ Elevation Head to provide adequate pressure to the community.

Pressure for community water use can be provided adequately by Pressure Head alone (standpipe), or by a combination of Pressure Head and Elevation Head (elevated tower).

When considering Elevation Head for calculations, a point of reference, or DATUM, is always chosen, from which to measure. It can be the gage at the tank bottom, or a point on the pipeline, but the datum can also be at some unconnected point beneath the system. Pressure Head and Elevation Head can still be determined from this point, even though in this case the Elevation Head is just potential energy that is being measured, for no pipeline extends down to the datum.

Perhaps the most practical reason for considering Elevation Head as a separate entity, is for comparison of two points along a pipeline. In determining the pressure differences between these two points, the pressure reading at Point A will be equivalent to the depth of the water in the tank (Pressure Head). The pressure reading at Point B will be equivalent to the depth of the water (Pressure Head) plus the vertical distance from Point A to Point B (Elevation Head).

Note: Elevation Head, when used in calculations, is always registered as feet of water. It is often symbolized as Z

Velocity Head

There is one more energy component in a water system: Velocity Head. Moving water, just as a rolling vehicle, possesses some energy because of its motion. For instance, it takes more energy to push a car from a standstill into motion, than it does to keep it in motion. The car has some energy because it is already moving. The same applies to water which has kinetic energy because of its velocity - Velocity Head. It is the distance which the water can move due to velocity energy. The formula for Velocity Head calculation is:

$$\text{Velocity Head} \ = \ \frac{\text{Velocity}^2}{2 \text{ x gravity}} \ = \ \frac{V^2}{2g} \ = \ \frac{V^2}{64.4}$$

For derivation see Appendix IV.

Note: Velocity Head, when used in calculations is always registered as feet of water.

In a pressurized pipe system, Velocity Head is usually the smallest of the three energy components. Normal water velocities range from 2 to 10 ft./sec. At a water velocity of 5 ft./sec., the Velocity Head would be .39 ft. Compared to Pressure Head and Elevation Head, this can be insignificant and calculations for it are sometimes omitted. In certain instances, however, it can be very important. Where water exits freely from the end of a pipe, there is no Pressure Head left, but the water is still moving. The energy keeping it going is its Velocity Head.

Where the flow is in an open channel: an aqueduct, stream, or partially filled sewer pipe, the water is not under pressure. It has only the downward slope of the channel to maintain velocity, and Velocity Head is the energy involved.

The most useful function of Velocity Head calculations in pressurized pipe systems is its potential as a conversion unit for flow calculation. Pressure Head and Elevation Head are registered on a pressure gage. Velocity Head is not. Recordable water pressure is transferred directly to the gage,

Pitot Gauge

because pressure in the pipe registers upward into the gage, as well as downward and to each side (Pascal's Principle). However, Velocity Head only registers forward, in the direction that the water is moving, and will not push upward into a pressure gage. A pitot gage, a flow measuring device, is used to capture Velocity Head, and compare it against Pressure Head. The device may have different configurations, but is basically composed of two gages installed as a unit. One of the gages has a bent end which is directed upstream into the flow. This one will record the pressure of the water (Pressure Head), plus the Velocity Head. The other gage extends straight into the pipe and registers only water pressure (Pressure Head).

To calculate flow with pitot gage:
- Convert psi readings from both gages to feet of water.
- Subtract one from the other; this value is the Velocity Head, in feet.
- Convert to water velocity with the Velocity Head formula.
- Use this and the cross sectional area of the pipe to calculate flow ($Q = AV$).

For practical use in field operations, flow dimensions can be set up on charts, based on these differential pressure readings.

Using this method, differential pressure readings will enable determination of flow. The real importance of Velocity Head in a system under pressure, is not its magnitude as an energy component, but its capacity to become the connecting link to flow calculations.

> **Note:** Velocity Head calculations are only performed in a dynamic water system. There is no Velocity Head in a static system.

If in the system pipe diameter remains the same from point to point, then the velocity is the same at both points, and the Velocity Head is the same at both points.

> **Note:** Velocity Head only changes from one point to another on a pipeline if the diameter of the pipe changes.

Refer back to the Hydraulic Grade Line as a designation of the total energy in the system. In a dynamic system, when Velocity Head is included, another grade line appears, which is called the Energy Grade Line. It is drawn just slightly above, and parallel to, the Hydraulic Grade Line. The numerical difference between the two is the Velocity Head.

total head

The total energy at any one point in a hydraulic system is the sum of the Pressure Head, Elevation Head, and the Velocity Head. The summation of these energies is expressed with a formula:

$$\text{Total Head} = \text{Pressure Head} + \text{Velocity Head} + \text{Elevation Head}$$

$$\text{Th} = \text{Ph} + \text{Vh} + \text{Z}$$

> **Note:** All the heads must be registered in feet of water for calculation.

head loss

When considering water traveling in a dynamic system, some pressure is lost along the length of the pipe because of friction encountered in rubbing against the pipe wall. Turbulence occurs, and energy is lost as heat which then dissipates to the surroundings. The amount of loss depends upon system length, diameter, flow, and interior roughness of the pipe. This is Friction Head Loss. It cannot be recovered, and must be calculated in by the utility when considering minimum system pressures to be maintained.

steady state flow

At this point, it should be emphasized that most hydraulic calculations deal with a system in Steady State Flow. This means that the amount of flow crossing a given point does not change with time. It is assumed that flow and pressure from the source are constant, even though in actual water systems flow varies as water tanks fill and empty.

One important fact becomes apparent in a steady state flow situation. <u>The flow that enters the system, exits the system</u>. As long as there are no leaks or lateral exits diverting the water elsewhere, it will not disappear or accumulate inside the pipe (water is incompressible); what goes in must come out. If the pipe diameter changes, the velocity will change, but the flow stays the same (Q=AV). In a dynamic system, friction losses will occur, and pressure readings will drop along the length of the pipeline, but the flow remains constant. If head losses are suddenly increased, this will cause pressure to drop, and less flow will enter the entire pipe system; but once it has entered, it must exit at the same flow rate.

bernoulli's theorem

It was Daniel Bernoulli, a Swiss mathematician and physicist in the 1700's, who first developed the calculation for the total energy relationship from point to point in a steady state fluid system.

Bernoulli's Energy Equation holds that, ignoring friction losses, from point to point in a fluid system employing steady state flow, the total energy is the same at every point in the path of flow, and these energies are composed of Pressure Head, Velocity Head, and Elevation Head. If there is a reduction in the energy of any one form, there must be an equivalent increase in another.

The Total Head formula is now applied to a system in which two points are considered:

<u>Point 1</u>		<u>Point 2</u>
$Ph_1 + Vh_1 + Z_1$	$=$	$Ph_2 + Vh_2 + Z_2$

A typical application is that of a pipeline in which the elevations of the two points differ. Drawing a datum horizontally through Point 2 (it is best to draw the datum through the point of lowest elevation, to avoid the use of negative numbers). The pressure at Point 2 is higher than that at Point 1, but the elevation of Point 1 is greater, by exactly the same amount. Inserting the numbers into Bernoulli's formula, they produce an equality. The increased Pressure Head that registers on the gage at Point 2, has resulted from the elevation of Point 1. Note that since the pipe diameter is the same throughout the system, the Velocity Head at both points is the same.

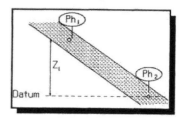

Take another example, that of a pipe whose diameter differs from point to point. Point 1 lies in the pipe with the narrower diameter, Point 2 in the wider pipe. Referring back to the Equation of Continuity (Q=AV), in a steady state system, if the cross sectional area increases, the velocity of the water will decrease. Therefore, the Velocity Head at this point will also decrease. Bernoulli's formula states that the total energy must remain constant from point to point. That means that one of the other forms of energy must increase at Point 2. Since this is a horizontal pipe, it is the Elevation Head. It is the Pressure Head that increases, and by the amount that the Velocity Head decreased from Point 1. When velocity decreases, pressure energy increases.

The opposite would happen if the pipe had decreased in diameter. Velocity increases, and pressure decreases. Utility personnel take advantage of this phenomenon when using the Venturi Meter, another flow measuring device. The structure of a Venturi is similar to a pipe which has been squeezed in the middle, creating a narrow diameter at the constricted section. A pressure gage is inserted at this point, and another one just upstream from the constriction. The gage at the constricted throat will read a slightly lower pressure than the one upstream, and the difference, when the readings are converted to feet, will be the difference in the Velocity Heads between the two sections. At normal water pipe flows, Velocity Head at the wider section often is assumed negligible, and so the pressure differential can be equated to the Velocity Head at the constriction. This is easily convertible to a value for velocity $(Vh = V^2/2g)$; then using the throat diameter and Q=AV, a flow value can be obtained.

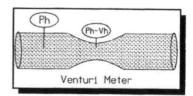

Venturi Meter

Note: When pressure changes because velocity changed, the difference between the two pressures is the <u>Velocity Head</u>.

It should be emphasized that these internal energy changes within a pipe system, are results of structural configuration, and can be reversed, by reverting back to the original structure. In this diagram, as in the Venturi Meter, the pipe constricts, and the pressure drops. The constricted section then enlarges to the original diameter, and the pressure at Point 1 is returned at Point 3.

Bernoulli's Formula has stated the theoretical equivalence of energies from point to point in a steady state flow system. However, in this form it does not account for the friction losses which occur in an actual dynamic system.

If the total energy at Point 1 is not equal to the total energy at Point 2 because there has been some pressure lost due to friction (energy at 2 is less), then it can be made equal by adding the Head Loss to the energy at Point 2.

The formula now becomes:

$$Ph_1 + Vh_1 + Z_1 = Ph_2 + Vh_2 + Z_2 + HL_f$$

As applied to real pipe systems, and as a mathematical formula, the energies at the two points are now equal.

Note: All quantities must be entered as <u>feet of water</u>.

There is another situation that should be pointed out here. When the system under consideration ends with an open pipe from which the water flows freely, all the pressure energy is reduced to zero as the water reaches the end of the pipe. The Velocity Head remains, however, and it keeps the water moving to exit the pipe. In other words, the pressure in a system is enough to overcome friction losses so that the velocity can be maintained. At the very end of the system,

friction losses have been overcome, and the pressure is reduced to zero, but velocity and flow remain. The flow admitted into and through the pipe is the amount that will reduce the head loss to zero at the end of the pipe.

In the case of a water main break, a great deal of flow emerges from the end of the pipe. Most of the Head Loss components are eliminated, but the system still has the same amount of source pressure, and all of that pressure must be depleted by the time the water gets to the end of the pipe. Flow increases enough to increase the head loss in each foot of pipe, so that all the pressure is gone as the water reaches the rupture.

Let us consider these three diagrams to the right. They could depict a pipe valved at the end, or a sink faucet, or a garden hose.

In I, the closed valve keeps the Hydraulic Grade Line horizontal. The Energy Grade Line is superimposed upon it; they are both at the same level. There is no Velocity Head, no velocity, and no flow.

In II, the valve is cracked open slightly. Flow is sacrificed to maintain pressure in the pipe, and the Hydraulic Grade Line drops only slightly up to the valve. Friction losses are small because flow is low. The Energy Grade Line follows parallel to it, and slightly above, for Velocity Head (and velocity) is small. As the water encounters the nearly closed valve pressure energy is converted to velocity energy at the narrow valve opening. Substantial friction losses also occur here because of turbulence in passing through the valve. The water pressure on the downstream side of the valve is reduced to zero, the only energy remaining is Velocity Head and there is free flow out of the pipe.

In III, the valve is fully open. Great flow is allowed through. Velocity is high, and the Energy Grade Line shows significantly above the Hydraulic Grade Line. Because of the greater velocity and flow, there is greater turbulence for every foot of pipe, and the Hydraulic Grade Line decreases steeply and evenly along the length of pipe, to the end, where the pressure is zero and the Velocity Head remains to keep the water flowing.

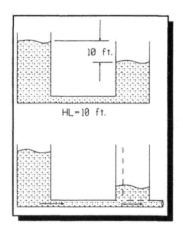

One other situation should be considered at this time. When a hydraulic diagram depicts one reservoir emptying into another, and points out the differences in water levels between the two, it is meant to bring out the fact that the decreased depth of water in the second reservoir is a direct result of the head loss encountered in the pipe as the water passed from one reservoir to the other. To make this clearer, adjust the diagram by extending the pipeline further to the right, and narrowing in the width of the second reservoir. The second reservoir (or any water tower) is nothing more than a fat piezometer, registering the pressure at that point in the pipeline.

If a diagram like the one described above were to be taken literally as drawn, and water were flowing from one to the other, the flow would gradually decrease as the second reservoir filled. The head loss would also gradually decrease, because the flow was decreasing. Eventually both flow and head loss are zero, and the surface levels in the two reservoirs are the same. But this is not a steady state flow situation. It is real, but it is not steady state, and steady state flow must be maintained for calculation.

For simple hydraulic situations, such as two points which differ in elevation alone, a quick view of the diagram allows mental calculation for the unknown. The solution would work out just as accurately if all quantities were inserted into Bernoulli's formula, but this is not always necessary. In more complex instances, however, such as a pipe system which changes in elevation, and pipe diameter, and involves some friction loss, the usefulness of having a well defined formula into which values can be inserted is obvious, in order to avoid mistakes.

Note: The secret to success is: DRAW THE DIAGRAM!

PROBLEMS

1. A flow of 2000 gpm takes place in a 12 inch pipe. Calculate the velocity head.

2. If water is flowing in a horizontal line at ground level from a 250 ft. elevated tank which holds water 20 ft. deep, at a velocity of 3.5 ft./sec., what are the:
   A. Pressure Head
   B. Velocity Head
   C. Elevation Head
   D. Total Head

3. A fluid with a Specific Gravity of 1.2 flows at 5.08 ml./sec. in a 4 inch line under a pressure of 100 psi. What is the:
   A. Pressure Head (ft. of water)
   B. Velocity Head in this line, in feet of water

4. What is the Velocity Head in a 6 inch pipe connected to a 1 ft. pipe, if the flow in the larger pipe is 1.42 cfs?

5. A 14 inch diameter horizontal pipe under a pressure of 120 psi has a water velocity of 20 ft./sec. The pipe then constricts to a 6 inch diameter. Assuming no friction head loss, what is the pressure in the 6 inch diameter section (psi)?

6. The Total Head in a six inch diameter pipe at a point 13.5 ft. above a datum is 19.3 ft. If the pressure at that point is 312 lb./sq.ft., what is the flow (cfs)?

7. An 8 inch diameter pipe reduces to a 4 inch diameter pipe. Pressure in the 8 inch pipe is 50 psi. In the 4 inch pipe the pressure is 10 psi. Find the flow. (assume no friction loss)

8. Velocity Head at a point in a 6 inch diameter pipeline is .44 ft. What is the flow (cfs) passing through the line?

9. A horizontal pipeline carries water at 30 ft./sec. At a point on the line the pressure is 20 psi. At another point, on an expanded section of this line, a gage registers 24 psi. What is the water velocity in the expanded section? (Assume negligible head loss.)

10. Assume negligible head loss.
    A. Calculate the velocity in a 6 inch diameter pipe discharging 1.4 cfs to atmosphere.
    B. Calculate the velocity of this discharge through a 2.5 inch diameter nozzle at the end of the pipe.
    C. Calculate the pressure intensity (psi) in the 6 inch pipe just upstream from the nozzle.

11. An industrial water treatment process employs multi-media pressure filters for turbidity removal. Process water is pumped in an upflow mode through the filters, which have a media depth of 4 ft. The pressure loss created by turbulence and friction in passing through the media is 8 ft. If the gage reading on the 6 inch diameter influent line is 65 psi, what is the gage reading on the 6 inch diameter effluent line?

12. Water is carried from a tank holding a 25 ft. depth, through a 12 inch diameter pipe. The water flows free at 25 cfs at the end of the pipe. What is the head loss due to friction?

13. A customer is watering his garden with a 1 inch diameter hose. Water pressure at the faucet is 50 psi. Head loss through the hose is 4 ft. What is the pressure of the water which is being discharged from the end of the hose?

14. A ground level storage reservoir registers a pressure of 13 psi at the bottom. It services a community 100 ft. below the tank in elevation. If the total head is 131 ft., what is the velocity of the water in the pipe (assume no head loss)?

15. A 6 inch diameter pipe is slanted upward and discharges water into the atmosphere. The flow rate is 1.963 cfs and the pressure at a point in the pipe 6 ft. below the discharge end is 33.5 psi. Calculate the total head loss between that point and the discharge end.

16. A high service pump will lift water 90 ft. into an elevated tank. The gage on the pump discharge line reads 45 psi. What is the head loss due to friction (ft.)?

17. A pump is moving water from reservoir #1, which is 40 ft. in diameter and full at 20 ft. water depth. The bottom of the reservoir is 100 ft. above the centerline of the pump. The water is pumped to reservoir #2 whose top is 200 ft. above the pump centerline. Both are cylindrical tanks of the same size.
    A. How much head must the pump provide (ft.) to fill reservoir #2 (negligible head loss).
    B. How many gallons will reservoir #2 hold when full?

18. If the water pressure lost in traveling from the storage tank, which stands 50 ft. off the ground, to the new housing development is 43.3 psi how many feet of water in the tank should be provided, in order to service the housing development with a minimum 30 psi pressure?

19. At a point in a 72 inch pipe the pressure is 60 psi, and the velocity is 2 ft./sec. This pipe reduces to 36 inches, and at a second point in this line, 10 ft. above the first, the pressure is 40 psi. What is the head loss from point one to point two?

20. At a point in an 8 inch diameter pipe which has an elevation of 690 ft., the pressure is 48 psi and the velocity head is .63 ft. If the headloss to a point downstream is 4.3 inches in a 10 inch diameter pipe at an elevation of 683 ft. what is the pressure in the 10 inch pipe?

21. A pitot gage attached to a 4" diameter pipe at a valve box registered a pressure differential of 4 psi on the gage. What was the flow?

22. A pump takes water through an 8 inch suction pipe and delivers it to a 6 inch discharge pipe in which the velocity is 8 ft./sec. In the suction pipe the pressure is -6 psi. In the discharge pipe, at a point 8 ft. above the pump, it is 60 psi.
    A. What is the pumped flow?
    B. Assuming neglig. HL, how many feet of head must the pump provide?

23. A 6 inch diameter pipe directs a jet of water vertically up into the air. If the rate of flow is 2000 gpm, how high will it go?

24. A fire pump delivers water through a 6 inch main to a hydrant. A 3 inch diameter hose is connected to the hydrant. It terminates in a 1 inch diameter nozzle. The nozzle is 10 ft. above the hydrant and 60 ft. above the pump. If friction losses are 28 ft. from pump to base of nozzle, neglecting air resistance, what gage pressure at the pump discharge is necessary to throw a stream of water 80 ft. vertically above the nozzle?

25. Water is siphoned from a tank at the rate of 20 cfs. The flowing end of the siphon is 214 ft. below the water surface. There is 1.5 ft. of friction headloss from tank to summit of siphon, and 1 Velocity Head is lost in friction from summit of siphon to end of siphon. The summit is 5 ft. above the water surface.
    A. Find the pipe diameter.
    B. What is the pressure at the summit?

CHAPTER 5

PUMPING - INTRODUCTION

pumping heads

A pump is installed to move water when existing pressure is inadequate. For instance, water will not move from the treatment plant up into the elevated storage tank on its own. The High Lift pumps provide the needed pressure to lift the water to this height. In determining the amount of pressure a pump must provide to get the job done, the nomenclature of pump heads was established.

Water enters the pump on the suction side, where the pressure is lower. Since the function of the pump is to add pressure to the system, discharge pressure will always be higher. In pump systems, measurements are taken from the point of reference to the centerline of the pump (horizontal line drawn through center of pump). Recall the distinction between:

Static Water System - water not moving

Dynamic Water System - water moving

Static Heads

The vertical distance from the water surface on the suction side, to the pump centerline, is the Static Suction Head (SSH). Static Head is always a positive value. It is pressure provided by the system, work the pump does not have to do. Even without the pump, the system would lift the water to this height on the right side of the diagram.

STATIC SYSTEM - Pump is Off!

Static Suction Head is registered on the gage at the suction side of the pump - when the pump is off (static system).

The vertical distance from the pump centerline to the water level on the discharge side of the pump is the <u>Static Discharge Head</u> (SDH). This is theoretically the total amount of work done by pump and system. The water was actually lifted this distance. The gage on the discharge side of the pump in a static system will register Static Discharge Head.

The vertical difference between the Static Suction Head and the Static Discharge Head is called <u>Total Static Head</u> (TSH). This is the theoretical amount of work that the pump must do in lifting water up to the higher tank. To provide this much head is why this pump has been installed.

STATIC SYSTEM - <u>Pump is Off!</u>

Pumps are frequently employed to draw water up from below the level of the pump, as in the case of a well. The configuration is such that on the suction side there is a lift, a <u>Static Suction Lift</u> (SSL). The pump must have the energy to provide this lift. On the discharge side, the pump is also providing the head needed to raise the water to the tank above, the <u>Static Discharge Head</u> (SDH). In pump systems which provide a suction lift, the <u>Total Static Head</u> (TSH) is the sum of the Static Suction Lift and the Static Discharge Head. The pump is doing <u>all</u> the work.

Dynamic Heads

If a pump system is to provide pressure to move water, obviously the pump must be on. Suction and discharge heads are not static. They are dynamic heads, and they will depend upon the length and condition of the system; suction and discharge heads will decrease with pipe length. The Hydraulic Grade Line will slope downward toward the pump on the suction side, and downward away from the

DYNAMIC SYSTEM - <u>Pump is On!</u>

pump on the discharge side, designating that some pressure has been lost from friction because this water is moving; the system is dynamic.

The <u>Dynamic Suction Head</u> (DSH) is the vertical distance between the water level on the suction side and the centerline of the pump, <u>minus the friction headloss</u>.

The <u>Dynamic Discharge Head</u> (DDH) is the vertical distance between the pump centerline and the water level on the discharge side, <u>plus the friction headloss</u>. The pump must make up this pressure lost in transit in order to get the water up into the tank. The Dynamic Discharge Head is the pressure that the pump discharge gage will record when the pump is on.

The <u>Total Dynamic Head</u> (TDH) is the difference between the Dynamic Suction Head and the Dynamic Discharge Head. The actual work this pump has to do in lifting water to the higher tank is greater than that designated by static heads. The pump must make up friction losses on both suction and discharge sides.

Again consider the case of a suction lift (below). Since pressure is lost in transit to the pump, the <u>Dynamic Suction Lift</u> (DSL) is the vertical distance between the water level and the centerline of the pump, <u>plus the friction headloss</u>. It will register as negative pressure (vacuum) on a gage on the suction side of the pump.

The <u>Dynamic Discharge Head</u> (DDH) is the vertical distance between the pump centerline and the water level on the discharge side, <u>plus the friction headloss</u>. The <u>Total Dynamic Head</u> (TDH) in this system is the sum of the Dynamic Suction Lift and the Dynamic Discharge Head - and it includes friction losses on both sides of the pump. This is the

DYNAMIC SYSTEM - <u>Pump is On!</u>

energy needed to move the water. It is this pumping head which determines the size of pump needed for this operation.

The same pumping heads can be identified in a booster pumping situation. When line pressure available from the source (an elevated tank, perhaps) has dropped too low, a booster pump is installed to increase pressure. There is no change in elevation; the pump

is installed solely to restore the pressure depleted by friction losses in the system.

Dynamic Suction Head is the line pressure on the suction side of the booster pump. Dynamic Discharge Head is the line pressure on the discharge side of the pump. Total Dynamic Head is the pressure that the booster pump is providing.

Note: Total Dynamic Head is the most important of the pumping heads. It is the head which is read from a pump curve.

---

power calculations

In order to calculate the cost of operating a pump, it is necessary to equate the amount of water moved to units of electrical energy consumed in moving that water. In order to purchase a pump, it must first be determined how much work that pump will be required to do.

Start with the concept of <u>work</u>, the operation of a force over a specific distance. The unit of work is the <u>foot pound</u>, the amount of work required to lift a one pound object one foot off the ground (ft.lb.).

When considering work being done, it is more valuable to be able to do the work faster, and so we consider the rate at which work is being done. This is called <u>Power,</u> and is labelled as <u>foot pound/second</u>. At some point it was determined that the ideal work animal, the horse, could move 550 pounds one foot, in one second. Since large amounts of work are also to be considered, this unit became known as <u>Horsepower</u>.

$$550 \text{ ft.lb./sec.} = 1 \text{ Horsepower}$$
$$\text{or}$$
$$33,000 \text{ ft.lb./min.} = 1 \text{ Horsepower}$$

A pump also does work, and has power. It lifts water (which has weight) a given distance, in a specific amount of time. The unit label is the same - ft.lb./min. Change the lb./min. to gal./min. (8.34 lb./gal.) and the formula states:

$$\frac{\text{ft. x gpm}}{3960} = \text{whp}$$

For Derivation see Appendix IV.

This is the standard formula for horsepower, derived from the amount of lift, or head, a pump will be expected to provide (in ft.) and the flow at which the pump will be operating (gpm). This is <u>Water Horsepower</u> (WHP), the power necessary to move the water to this height.

A pump, however does not operate alone. It is driven by the motor drives, and electrical energy drives the motor. Neither pump nor motor are 100% efficient. There are friction losses even within both of these units, and it will take <u>more</u> horsepower applied to the pump to get the required amount of horsepower to move the water, and even <u>more</u> horsepower applied to the motor to get the job done.

<u>Brake Horsepower</u> is the horsepower applied to the pump.

<u>Motor Horsepower</u> is the horsepower applied to the motor.

<u>Efficiency</u> is the power produced by the unit, divided by the power used in operating the unit.

Water Horsepower must be divided by the pump efficiency to obtain the Brake Horsepower. Brake Horsepower must be divided by the motor efficiency to obtain the Motor Horsepower

Most pumps operate at 60-85% efficiency. Motors are usually 80-90% efficient.

Multiplying the pump efficiency by the motor efficiency will provide the combined, or total efficiency of the system, the <u>Wire to Water Efficiency</u>.

Next, it is necessary to be able to convert power from units of Motor Horsepower, to units of electrical energy (kilowatts). There is a direct conversion:

$$1 \text{ Horsepower} = .746 \text{ Kilowatts}$$

This power demand, in Kilowatts, is sold by the hour (Kilowatt hours). Multiply then, by the number of hours the pump runs, and then by the cost per Kilowatt Hour, to obtain the cost of running the pump.

In summary:

$$\frac{gpm \times head}{3960} = \frac{water\ Hp}{pump\ effic.} = \frac{brake\ Hp}{motor\ effic.} = motor\ Hp \times .746\ \frac{Kw}{Hp} \times hrs. \times \frac{cost}{Kw.hr.} = \$$$

well problems - pumping problems

These are basic calculations which apply to water pumped from a groundwater aquifer, as well as to water pumped from a wastewater wet well or tank. The principles, terminology, and calculations apply to both.

Groundwater Well

Well Yield: the volume of water pumped from a well in a specified time period. Well yield is expressed as gallons per minute (gpm).

Static Water Level: the water level when the pump is off.

Pumping Water Level: the water level when the pump is on.

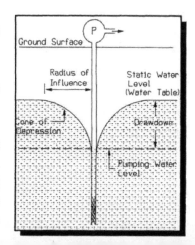

Drawdown: the difference, measured in feet, between the Static Water Level and the Pumping Water Level. It is the distance the water level drops once pumping begins. The vacuum created at the end of the pump intake line brings water from the immediate area into the well, and a depression is caused in the water level in that area. It is important to note that once drawdown exists, the pump chosen must be powerful enough to lift the water that extra distance, and still provide the needed capacity.

Safe Well Yield: the maximum pumping rate that can be achieved without increasing the drawdown. In a groundwater well, the Drawdown may easily be fifty feet or more. It may take an hour or so to stabilize and arrive at a permanent Pumping Water Level, but once this is achieved, the Drawdown should not be increasing with time. If the Drawdown is 20 ft. this month, it should be 20 ft. next month, and next year, regardless of whether that pump is run continuously, or in an on/off mode. If the Drawdown increases as time passes, the pump is withdrawing more water than the aquifer can return to the well. This is called Water Mining. Eventually, the Pumping Water Level will drop below the end of the well screen, and the well will fail.

Overpumping an aquifer can have permanent effects on the land itself: sinkhole formation, and in areas proximate to the ocean, saltwater intrusion. Natural phenomena can also influence well supply. Drought lowers the Static Water Level, so that even with acceptable Drawdown the Pumping Water Level may be too low in dry years for effective pumping (many small private wells run dry during times of drought). Excessive rainfall raises the Static Water Level, reversing this effect. Soil porosity and geologic structure also influence the aquifer's ability to return groundwater to the well, and limit pump size.

Radius of Influence: the distance between the pump shaft and the outermost area affected by drawdown. This becomes important in well fields with many pumps. If wells are set too close together, the Radii of Influence will overlap, increasing the drawdown in all wells. Pumps should be spaced far enough apart so that this does not happen.

Cone of Depression: the curve of the line which extends from the Pumping Water Level to the Static Water Level at the outside edge of the Radius of Influence. Design calculations based on the slope of this curve will determine the Radius of Influence.

Specific Capacity: pumping rate per foot of Drawdown (gpm./ft.). This is a measure of Well Yield. Given the same pumping rate, if the Specific Capacity is decreasing with time, the Drawdown is increasing.

Wet Wells

Water pumped from a tank by a pump set above the water surface exhibits the same phenomena as the groundwater well. There is a slight depression of the water surface right at the intake line (drawdown) while the pump is operating, but in this case it is minimal because there is no restriction of the water approaching the pump entrance. As the pump draws water up, it fills into that space almost immediately, so that Pumping Water Level is barely lower then Static Water Level. Most important at this location is to ascertain that the suction line is submerged far enough below the surface, so that air entrained by the active movement of the water at this section is not able to enter the pump.

If the pump is operating at exactly the same rate as the tank is filling and is on continuously, then the water level in the tank will remain the same. However, for water and wastewater applications, this is very rarely the case. Water enters the tank at varying flows, depending on the source, and since wet well size is limited, pumping ability must accommodate for large and small flows, and will almost never exactly match the inflow. To cope with this, variable speed pumps can be used, or several pumps are installed for single or combined operation. In some cases pumping is done in an on/off mode, as with sump pumps. Control of pump operation is in response to water level in the well. Level control devices will sense a high and low level in the well, and transmit the signal to the pumps for action. In wet well calculations, any lowering of the water level in the well is referred to as Drawdown. A common method for calculating pump performance (pumping rate) is to run a pumping test. Water is pumped out of a tank of known volume, drop in water level is recorded (drawdown), and that volume of water is converted to gallons. Divide by the time span of the pump test, and this will yield gallons per minute pumping rate. Note that if there is a simultaneous inflow to the well, this will have to be calculated also. See Appendix III.

A water meter may be installed on the pump discharge line to monitor pumping rate, but a pumping test can be useful to check the accuracy of the meter.

water level sensing and pump control

Accurate sensing of high and low water levels in a wet well is imperative for proper timing of pump operation. In groundwater well maintenance, accurate recording of Static Water Level, Pumping Water Level, and Drawdown, is just as important.

There are several different types of water level sensing devices in common use:

Mechanical Float System

In use for many years, this system is installed in tanks and wet wells to sense high and low water levels for signaling pump operation. Steel rods have attached floats which extend from the side of the tank and ride up and down on the water surface. At maximum and minimum level the system mechanically activates a transmitter which signals the pump to action. These systems are bulky, but simple in design and reliable.

Diaphragm Element

This system, also in use with tanks, operates with a small air-filled chamber. The bottom section of it is composed of a diaphragm, flexed outward when the water level is low. The level rises, and at a specific depth the pressure generated by the water depth pushes the diaphragm inward, compressing the volume of air inside. This increased air pressure is registered on a sensor and transmitted to activate the pump. This type of system is often installed into the bottom of sump pumps; the water level rising over the pump switches the pump into operation. As the water level goes down, pressure decreases, the diaphragm flexes outward again, and the pump switches off.

Bubbler System

Used in water level sensing in wet wells, as well as for reading levels in groundwater wells. It requires a source of air (compressor or tire pump) which is pumped through a narrow copper tube extending to the bottom of the well or tank. With the air source turned off, the water will fill the tube to the level of the surface. Air pumped into the tube expels the water in it. As soon as all the water is evacuated from the tube, air bubbles out the bottom. A pressure gage at the top of the airline records the pressure needed to accomplish this, which, in feet, is equivalent to the water depth in the tank or well. If a pressure reading is recorded when the well pump is off (Static Water Level), and when the pump is on (Pumping Water Level) - the difference between these readings is the Drawdown. Wastewater lift stations commonly employ this system to signal high and low water levels for pump operation.

Electronic/Probes Transducers

Electronic probes are installed at known depths and sense water rising over them. Transducers convert the signal, making gas pressure in the line correspond to the water pressure in the well. Care must be taken to avoid grease accumulations on the probes.

Ultrasonic Detectors

A transmitter is mounted over the wet well, pointed vertically downward. High frequency sound waves are sent to the water surface; they bounce back, and the time span is measured. This is electronically converted to a milliamp signal which is proportional to the time, and to the water level. This system is in popular use in wastewater treatment, for no part of the detector touches the water.

PROBLEMS

1. Water is pumped at 500 gpm to a water storage tank on the roof of an industry. The gage on the pump discharge line reads 95 psi. The difference in elevation between the gage reading and the water level in the tank is 90 ft. What is the pressure loss (ft.) in the piping?

2. A wet well pump operating with a suction lift has a Total Dynamic Head of 150 ft. If the Dynamic Discharge Head is 130 ft., and the friction head loss on the suction side is 10 ft., what is the vertical distance between the water level on the suction side and the centerline of the pump?

3. A flow of 1840 gpm is to be pumped against a head of 40 ft. What is the water horsepower required?

4. A 12 hp pump delivers 1200 gpm. Assuming the pump is 100% efficient, what is the pressure (psi) against which the pump is operating?

5. A pump is putting out 5 whp and delivering a flow of 430 gpm. Assuming no friction head loss, what is the height to which this pump is lifting water?

6. A pump is 75% efficient, and the motor is 82% efficient. If the pump is delivering 825 gpm against a Total Dynamic Head of 95 ft.:
    A. What is the Water Horsepower?
    B. What is the Brake Horsepower?
    C. What is the Motor Horsepower?

7. If the motor is 94% efficient and the pump is 90% efficient, what horsepower will be required to deliver 200 gpm against a total pressure of 45 psi?

8. A motor draws 2238 Watts. The weekly charge for operating the pump is $1.65. If the cost per Kilowatt Hour is $.015, how many hours per week is the pump in operation?

9. A 25 hp pump is operating against a total pressure of 30 psi. If the efficiency of the pump is 88%, what is the maximum gpm delivery?

10. A pump 86% efficient delivers 590 gpm with a Brake Horsepower of 6.5 hp. What is the suction lift in feet if all other heads are equivalent to 34 ft.?

11. A 5 hp pump/motor combination is pumping 355 gpm against a Total Dynamic Head of 41 ft. If the motor is 91% efficient, what is the pump efficiency?

12. A pump is discharging 700 gpm against a head of 82 ft. The wire to water efficiency is 83% and the power rate is $.08/Kw.hr. What is the power cost for 14 hours of pumping?

13. A 300 gpm centrifugal pump is operating against 15 ft. of lift and losses on the suction side, and a discharge pressure of 50 psi. Instruments placed in the motor circuit show that it consumes 5.5 Kw.hr. in 30 minutes. What is the wire to water efficiency?

14. Water is pumped at 350 gpm from a well to a storage tank 400 ft. above. If the pump system's wire to water efficiency is 72%, how much power in Kw does the well require?

15. A water level gage on a 500 gpm well pump registers 52 psi when the pump is off. When the pump is running, this gage registers 39 psi.
    A. What is the Drawdown?
    B. What is the Specific Capacity of the well?

16. Given the following information on a well:
    Static Water Level: 400 ft.
    Casing diameter: 12 inches
    Well depth: 1200 ft.
    Pumping rate: 150 gpm
    Pumping Water Level: 500 ft.
    A. What is the Drawdown when the pump is operating?
    B. What is the Specific Capacity?

17. Checking water level in a well: when 400 gpm is being pumped the level gage pressure reading is 29.5 psi and the Drawdown is 52 ft. When the pump is shut off the water level rises to 200 ft. below ground surface. What is the pressure reading at that time (psi)?

18. A water level gage 200 ft. long extends from the ground surface to the bottom of a well. With the well pump off, the line is pressurized and a pressure of 55 psi is read on the gage. Well pump is then turned on, and after stabilizing, 32 psi is read.
    A. What is the depth to the water table?
    B. What is the Drawdown?

19. A tank 20 ft. in diameter and 10 ft. deep is half full. Flow entering equals 100 gpm. Pumping out 50 gpm. How long does it take to fill the tank?

20. A wastewater wet well is 10 ft. in diameter and 9 ft. deep. With no pump operating, the rise in water level is 2 ft. in 1 min. 46 sec. A pump is started and the water level begins to drop at a rate of 4 inches in 4 minutes. What is the gpm pumping rate?

21. During a 30 minute pumping test, 3680 gallons are pumped into a tank which has a diameter of 10 ft. The water level before the pumping test was 3 ft. What is the pumping rate in gpm?

22. Water is pumped into an empty 22 ft. diameter tank at the rate of 200 gpm. At the end of 4 hours of pumping, what is the gage reading at the bottom of the tank (psi)?

23. A community's 75,000 gallon elevated storage tank is one third full. The wells are pumping at a steady rate of 500 gpm. The distribution system demand is constant at 200 gpm. How much time will elapse before the tank is full (hrs.)?

24. A lift station wet well is pumped at a rate of 150 gpm for 20 minutes each hour. A constant flow of water is coming into the station at a rate of .1 cfs. What is the problem with this station, and offer a positive solution for it.

25. A wet well is 11 ft. by 11 ft. by 10 ft. 6 inches deep. The influent rate is 110 cfm, and when the pump is operating, you measure a drop in the water level of 15 inches in 4 min. 20 sec. If the pump is designed for 1200 gpm delivery, what percent of design efficiency is the pump operating at?

CHAPTER 6

FRICTION LOSS

This chapter will consider friction head loss in closed pipe systems. The work is straightforward once the components are understood and inherent weaknesses in theoretical calculation are overcome by adaptation to each particular water system.

When water flows in a pipeline under pressure, the forward velocity is produced by the weight of water behind it. It does not matter whether the pressure is originating from a standing tank of water (Pressure Head), or from the slope of the pipe (Elevation Head), or from a combination of the two.

In a dynamic system, pressure decreases along the length of the pipeline; the Hydraulic Grade Line slopes downward, designating pressure loss from friction created as the water encounters the sides of the pipe. This is <u>Major Head Loss</u>, and usually accounts for most of the pressure drop in dynamic water systems. <u>Minor Head Loss</u> is caused by extra turbulence created at bends, fittings, and diameter changes in the pipeline, and in most systems is a lesser component of pressure loss. Minor Head Loss will be considered in Chapter 8.

Since the source of head must provide the needed lift and overcome losses, calculations for head loss are important considerations even before a system is developed, and prior to new construction, pipe replacements, and expansions.

components of head loss
Roughness
The interior surface of a pipe is rough, even when the pipe is new. This roughness is dependent upon pipe material, and will increase with corrosion and age. Normal flow

in water pipes is turbulent, and that turbulence increases with pipe roughness; energy is spent, and pressure drops over length.

Length

Friction losses occur with every foot of pipe length, and this must be included in head loss calculations. The longer the pipe, the more head loss.

Diameter

Pipe diameter is also an important factor. In a large diameter pipe, less of the water is actually touching the sides of the pipe and encountering friction, than in a small diameter pipe. Therefore, small diameter pipes have more potential head loss than large diameter pipes.

Velocity

The faster the water flows through the pipe, the more turbulence is created, and the more head loss there will be. So the velocity, or more appropriately, the energy due to velocity, Velocity Head ($V^2/2g$), also has an effect. It can also be stated this way: for the same diameter pipe, as the flow increases, head loss increases ($Q=AV$).

In summary, Friction Head Loss (HL_f) is dependent upon:

- material and condition of pipe (roughness)
- length of pipe
- diameter of pipe
- Velocity Head

From these common sense components, a formula to calculate head loss was developed.

$$HL_f = f \times L \times \frac{1}{d} \times \frac{V^2}{2g}$$

This is the Darcy-Weisbach formula, and has been the basis for head loss calculations since the 19th century. Pipe roughness is designated by a fractional number, f, the friction factor, or Coefficient of Friction. The Darcy-Weisbach formula as such, was meant to apply to the flow of any fluid, and into this friction factor was incorporated not only the degree of roughness, but also an element called the Reynold's Number, which was based on the viscosity of the fluid, and the degree of turbulence of flow. Calculations for f are complex, and the formula for it varies, depending on whether the flow is laminar or turbulent. Special charts were developed to help ease calculations, but even they are difficult to interpret, and lose accuracy in systems approaching critical velocity (where laminar flow starts to become turbulent).

The Darcy-Weisbach formula was flexible enough to adapt to most fluids and velocities, but for popular use in water systems, where the flow is always turbulent, and where the damping effect of viscosity is negligible and unchanging, a more easily usable formula was developed by Hazen and Williams:

$$Q = .435 \times C \times d^{2.63} \times s^{.54}$$

For Derivation see Appendix IV.

Note: Pipe dimensions are in feet, flow is cu.ft./sec.

With a calculator it is easy to compute, and all components can be measured or estimated directly. The Hazen Williams formula also originates from the concepts of roughness, length, diameter and Velocity Head, the basic components of Head Loss.

With the Hazen-Williams formula came a new designation of pipe roughness. This is the C factor, the modern day Roughness Coefficient. C does not vary appreciably with velocity, and by comparing pipe types and ages, it includes only the concept of roughness, ignoring fluid viscosity and Reynold's Number.

Based on experimentation, accepted tables of C factors have been established for pipe. One is included here (see end of chapter), but there are other similar tables. Estimates of roughness coefficients may be made mentally for pipes not included, and generally C factor decreases by one with each year of pipe age. Flow for a newly designed system is often calculated with a C factor of 100, based on averaging it over the life of the pipe system. A knowledge of the particular system is invaluable here, for pipe roughness is very dependent upon water quality, and historical data on the chemical condition of the particular water and its effect on the interior surfaces of the pipes can determine what value is chosen for C.

Note: A high C factor means a smooth pipe. A low C factor means a rough pipe.

---

slope

"Slope" as an English word means the amount of slant, or incline. This applies to water systems also. In open channel systems (aqueducts, sewer pipe), where the water flows by gravity, slope is the amount of incline of the pipe, and is calculated as feet of drop per foot of pipe length (ft./ft.). Slope is designed to be just enough to overcome frictional losses, so that the velocity (and flow) will remain constantly, the water keeps flowing, and solids will not settle out along the way. The same principle holds true in a closed system under pressure. In a horizontal conduit which encounters "feet" of pressure loss for every foot of pipe, the pressure loss would slow down the velocity, restricting flow, and so must be overcome. It is the same as the slope in an open system, but instead of actually slanting the pipe to overcome the losses, a source of pressure is added to overcome friction. It is still called "slope": feet of head loss per foot of pipe (ft./ft.), which must be overcome. To calculate, divide pressure loss by length of pipe:

$$s = HL_f / L$$

The most frequently used version of Hazen-Williams formula is very similar:

$$Q = .435 \times C \times d^{2.63} \times s^{.54}$$

There are many variations of the Hazen-Williams formula. Some have slightly different exponents, some equate to velocity instead of to flow and some are specialized for open channel flow. Other hydraulics experts have developed their own variations, and use them under their own names, but it is easy to recognize the general format of the equation, and they all calculate out to yield the same approximate value of any one dimension. It is important to stress "approximate", for a few inherent weaknesses occur in any of these formulas.

There is no problem with flow, velocity, slope and length. If these are to be the known factors, they can be measured, metered, read from gages. The weakest point in the formula is the Roughness Coefficient, C. If this is to be a known factor for calculation, at best it can only be estimated, based on pipe type, age, and knowledge of water quality. The only way to accurately determine the amount of roughness in a pipe is to measure all the other components of the formula, and then calculate for C. This is done by utilities to determine when a pipe needs to be cleaned.

For example: Isolate a pipe section. Attach gages to two hydrants, one downstream from the other. Open a third hydrant downstream from both of these to induce flow. Use a pitot gage to measure flow at the open hydrant. Use the Hazen-Williams formula to calculate for C factor.

Pipe diameter can also lead to inaccurate calculations. Inside diameter is actually what should be used (that's where the water is), but even using rated, or nominal diameter, little accuracy is lost. Most important is knowledge of the particular system's water quality. How much rust buildup, chemical post-precipitation, or scale formation have occurred since that pipe was laid? A six inch diameter pipe can quickly become a two inch diameter pipe if the water quality is poor. In how many systems can we assume that the diameter of the pipe is the same as when it was new? This is where the operator's knowledge of his treatment system counts.

For field use, as an alternate to calculating the Hazen-Williams formula, an alignment chart has become popular, and can be used with reasonable accuracy. This chart is of a special type called a nomograph. It may yield an answer with one or more of its dimensions unknown. Dimensions of discharge, pipe diameter, water velocity, C factor, and head loss in ft./1000 ft are entered.

Values of C for Hazen-Williams Formula

Type of Pipe	C
Asbestos Cement	140
Brass	140
Brick Sewer	100
Cast Iron	
10 years old	110
20 years old	90
Ductile Iron, (cement lined)	140
Concrete or Concrete Lined	
Smooth, steel forms	140
Wooden forms	120
Rough	110
Copper	140
Fire Hose (rubber lined)	135
Galvanized Iron	120
Glass	140
Lead	130
Masonry Conduit	130
Plastic	150
Steel	
Coal-tar enamel lined	150
New Unlined	140
Riveted	110
Tin	130
Vitrified	120
Wood Stave	120

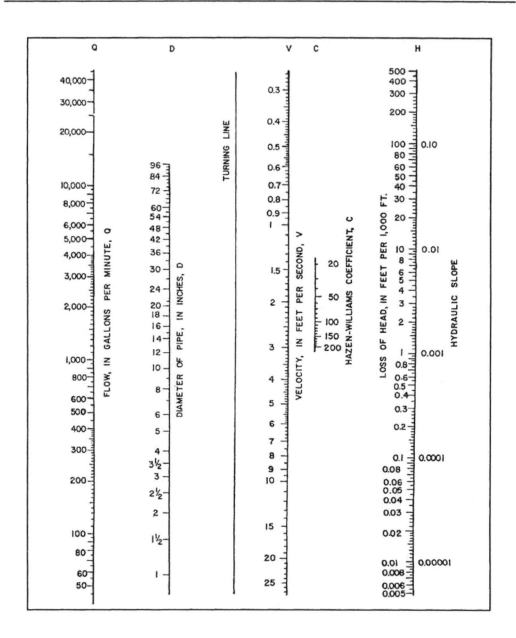

Hazen-Williams Alignment Chart

Directions for Using Hazen-Williams Alignment Chart

For determining head loss, find the proper discharge and line it up with the pipe diameter. Extend the line across to the Turning Line, and mark the point where the Turning Line was crossed. Line up that point with the proper C factor on the other side, and extend it to the head loss.

Velocity is determined from a line which passes through flow and diameter. Turning Line is not needed for this. It is needed only for finding C factor or head loss.

PROBLEMS

1. A 36 inch diameter pipe carries a flow of 800 gpm. If this pipe has a C value of 100, what is the slope (ft./ft.)?

2. A flow of 10 cfs occurs in a rough concrete pipe of 30 inch diameter. What is the total head loss due to friction in a 2000 ft. length of the pipe?

3. What is the head loss in 350 ft. of 10 inch diameter new unlined steel pipe carrying .8 MGD?

4. A 4 inch diameter cast iron main 40 years old and 1000 ft. long carries .2 cfs through a section of town. How much pressure (ft.) must be provided by the utility to maintain the flow for this section alone?

5. An elevated tank 50 ft. off the ground supplies a house with water. The water flows at 2 cfs through a 10 inch main 500 ft. long (10 year old CIP). How deep does the water have to be in the tank in order to provide the house with a pressure of 40 psi?

6. A 9000 ft. long, 18 inch diameter AC pipeline 10 years old delivers 4.5 MGD to a community which is 100 ft. higher in elevation than the treatment plant. The pressure delivered must be at least 60 psi. What pressure must be provided at the treatment plant to achieve this (psi)?

7. Water flows through a 6 inch diameter horizontal galvanized pipe 1000 ft. long with an average velocity of 8 ft./sec. What is the pressure drop (psi) across this line?

8. What minimum size lined steel pipe should be used to carry a flow of at least 10 cfs if the total friction loss is to be 25 ft. over each mile of pipe length?

9. In a test of 20 inch diameter cast iron pipe, the flow is steady at 2.2 cfs, and the Hydraulic Grade Line dropped 4 ft. in a length of 8000 ft. of pipe. What is the value of C?

10. A 10 inch diameter horizontal concrete pipe (C=120) carries water at 1500 gpm from point 1 to point 2. The distance between the two points is 1000 ft. The pressure at point 1 is 80 psi.
    A. What is the pressure at point 2?
    B. What is the difference in the water velocity between the two points?

11. Water flows through a 12 inch diameter galvanized iron pipe 800 ft. long with a velocity of 9 ft./sec. The pipe outlet is 50 ft. above the pipe inlet. What is the pressure difference between inlet and outlet (psi)?

12. What lost head per foot of pipe in a 16 inch diameter lead pipe will create the same flow as occurs in a 20 inch diameter (lead) pipe with a drop in the HGL of one foot per thousand feet?

13. Determine the discharge through a new concrete lined steel pipe of 1 ft. diameter if the pressure drop across the length of pipe is 2 psi per 1000 ft.?

14. When first installed between two reservoirs a 4 inch diameter cast iron pipe 6000 ft. long conveyed 1 cfs of water. If after 15 years, a chemical deposit had reduced the effective diameter to 3 inches, what then would be the flow rate? (assume no change in C value)

15. What percent error is introduced into Q when C is misjudged by 20%?

16. A. What size pipe would be required to transport 3900 gph of water if the head loss in 570 ft. of the masonry pipe had to be less than or equal to 1.18 ft.?
    B. If it was necessary to maintain at least a velocity of 2.5 ft./sec., how much head loss would be in the pipe?

17. Given the 800 ft. pipe below, calculate the friction head loss A-B(ft.), (C=100)

18. With 1 cfs flowing, what is the required total head to maintain flow at steady state in the system shown (assume 20 year old CIP)?

19. A sewer carries a full flow of 5.28 cfs. If C=100 and slope = .003 ft./ft., what is the water velocity?

20. A new PVC pipeline from a reservoir to a subdivision is 1.5 miles long and 8 inches in diameter. The reservoir surface elevation is 325 ft. A hydrant at the subdivision is at elevation 100 ft. To maintain at least 20 psi at the hydrant, what flow and velocity could be delivered from the hydrant?

21. A pressure gage at point 1 on a horizontal 8 inch diameter pipeline reads 45 psi. At point 2 on the same pipe, 100 ft. downstream, another pressure gage shows the same reading. Assuming that both gages are functioning correctly, what is the flow?

22. The system has 5000 ft. of 12 inch diameter pipe with a C factor of 110, carrying a flow of 1.7 MGD.
   A. What is the total head loss?
   B. If the pipe was replaced with a 16 inch diameter pipe, with the same C factor, at the same flow, what would be the pressure drop across this line (ft.)?

23. A horizontal high pressure line feeds water across an osmotic membrane at 400 psi. The head loss across the membrane is 700 ft. What is the pressure on the downstream side of the membrane (psi)?

24. Given the diagram to the right, find:
   A. Tank diameter.
   B. Condition of pipe (C factor).
   C. Gage pressure.

25. An old 14 inch diameter main (C=60) carries water 2000 ft. to a pond where the flow is free out the end of the pipe. A gage on the upstream end of the pipe reads 35 psi. The pipe is then cleaned and has a new smooth plastic liner inserted. Lining the pipe decreases the diameter to 13 inches.
   A. What was the flow through the old pipe?
   B. What is the flow through the new pipe?

CHAPTER 7

COMPOUND PIPES

Municipal water distribution systems are not built with separate single pipes extending from treatment plant to customer's residence. Rather, a network of pipes is laid under the streets - different diameters, lengths, ages and materials. Some branch and loop; some are connected end to end. These more complex arrangements are referred to as <u>Compound Pipe Systems</u>. There are two basic types:

<u>Pipes in Series</u> - two or more different pipes, laid end to end.

<u>Pipes in Parallel</u> - two or more pipes laid side by side, with the flow splitting among them.

Series Pipes   Parallel Pipes

calculating with pipes in series

When pipes are laid in series:
- A) The same flow passes through all of them.
- B) The total head loss is the sum of the head losses of all of the component pipes.

Situation I

In series systems where the flow is given, and the total head loss is the unknown:
1) Use the Hazen-Williams formula to solve for the slope and the head loss of each, as if they were separate pipes.
2) Add up the head losses to get the total head loss.

Situation II

A system where the total head loss is given and flow is the unknown is a little more complicated. Consider the system below:

Total HL_f = 50 ft.

C = 100 all pipes

Solve for Q

Using the Hazen-Williams formula:

$$Q = .435 \times C \times d^{2.63} \times s^{.54}$$

$$Q = .435 \times 100 \times ? \times ?$$

- Which diameter pipe should be used?
- In solving for slope, one would normally use the formula $s = HL/L$, but this head loss applies to both pipes together, and it is not known how much loss originates from each one.

In such cases, the <u>Equivalent Pipe Theory is used</u>.

One single <u>Equivalent Pipe</u> must be created which will carry the correct flow, because the head loss through it is the same as that in the actual system. The equivalent pipe can have any diameter and C factor, just as long as those same dimensions are maintained all the way through to the end. From the diagram, if the equivalent pipe is long enough to yield 50 ft. of head loss, then the flow passing through this pipe will be the same as the flow in the actual system.

Note: Whatever the flow and total head loss may be, they are the same in the actual system, and in the equivalent pipe.

This equivalent pipe must have the right length, so that it will allow the correct flow through, which yields the correct head loss (the given head loss).

1) Start by looking back at the actual system diagrammed. There must be a flow and a head loss for this system, which can be applied to the equivalent pipe, in order to solve for a length. <u>Assume</u> a flow -any flow.

2) Using Hazen-Williams formula, solve for slope, and head loss (based on the assumed flow) for each pipe component of it.

3) Add them up to get the total head loss through the system. This is not the true head loss (true head loss is 50 ft.), but it is correct for the assumed flow.

4) With the knowledge that flow and head loss are the same in the equivalent pipe as in the actual system, construct the equivalent pipe. Choose any diameter and C factor; use the assumed flow and solve for the slope.

5) Now solve for the length of the equivalent pipe. Using s=HL/L, insert the head loss just acquired from calculation with the assumed flow. Insert slope just acquired from calculation with the assumed flow. Solve for length.

6) Since both head loss and slope were derived from calculations based on the same assumed flow, they have a specific numerical relation to each other. This relation would be true, and would be the same, no matter what had been chosen as the flow. This relation - is the length of the equivalent pipe. It would be the same length no matter what flow had been used - even if it had been the true flow of the actual system. This length, therefore, is a <u>real</u> length!

7) Use it! Again, with s=HL/L, insert the length of the equivalent pipe. Go back to the actual system, and insert the head loss (given 50 ft.). Solve for slope.

8) Using Hazen-Williams formula, solve for the flow actually passing through that pipe, which is the flow of the original system.

Check your answer. Insert this calculated flow into Hazen-Williams formula for the actual system; solve for slope and head loss of each pipe component; add them up. It should come to 50 ft. total head loss.

Note: Equivalent pipes may be substituted for actual pipes in any system.

---

calculating with pipes in parallel

When pipes are laid in parallel:

A) The <u>flow splits</u> among them. Least flow is admitted to the pipe with the most resistance (smaller diameter, longer, rougher). The greatest flow will pass through the pipe with the least resistance (wider, shorter, smoother).

B) The <u>head loss</u> in each pipe component of a parallel system - <u>is the same</u>. The flow split makes it so. Taking the path of least resistance, just enough flow is admitted to each pipe, so that the head losses are balanced. Each pipe has the same head loss.

C) The head loss through the entire parallel system is the same as the head loss through any one of its pipe components. If one equivalent pipe were substituted for this entire system, and the approaching flow were passed through it, the head loss registered would be identical to that registered through each component pipe of the parallel system. That is what makes this pipe equivalent.

D) In parallel pipe systems, the <u>head losses are not added</u>. The flows are added.

Situation I

In parallel pipe systems, when the actual head loss is given (across one pipe segment, or across the whole system), and the incoming flow, as well as the component flows, are unknown:

1) Take one pipe of the set, divide head loss by length to get slope.

2) Use Hazen-Williams formula, solve for flow in that pipe.

3) Do the same with the other component pipes.

4) Add up the flows to get the incoming flow.

Situation II

In parallel systems where the approaching flow is given, but the flows and head losses of the components are to be determined:

1) The entire system can be replaced with one equivalent pipe.

2) Assume a head loss; using Hazen-Williams formula and the actual system, solve for flow in each pipe based on the assumed head loss.

3) Add up these flows to get total flow.

4) Create the equivalent pipe. Choose any diameter and C factor. Calculate what length pipe will give us this total flow for this assumed head loss (use Hazen-Williams formula and the new total flow to solve for slope - in the equivalent pipe).

5) Using slope for the equivalent pipe which is based on the assumed head loss, solve for equivalent pipe length (s=HL/L). This is the <u>real</u> length.

6) Now replace the parallel system with this one equivalent pipe. It has all the needed dimensions.

7) Take the actual flow incoming to the system. Solve for slope, and head loss, in the equivalent pipe. This head loss is the same as the head loss in each segment of the original parallel system.

8) To solve for flows in the separate pipe components of the parallel system, use this head loss to get slope, and then flow, for each.

To check the answer, add up the flows. They should sum up to the incoming flow which was originally given.

There is a short cut which may be used in a case like this. After proceeding through step 3, flows have been created for the actual system which were based on an assumed head loss, and so are not numerically true. However, the percentage flow split among the pipes is valid. Take each component flow derived from the assumed head loss, and calculate its percentage of the total flow. Multiply each percentage by the actual approaching flow to get the true component flows in the parallel system. To solve for head loss in each pipe, use Hazen-Williams formula to derive slope, and then head loss.

Note: In parallel pipe systems, head losses are <u>not</u> additive. Flows are additive.

The type of calculations covered in this chapter are the basis for solving piping network problems. An analysis of flows and pipe conditions in an entire section of a city water distribution system may be required for planning system expansions, maintenance, major replacements, and for troubleshooting low pressure problems.

There are several solution methods in use. The equivalent pipe method which has been chosen is one of the simplest. Another method of network analysis is the Hardy-Cross system. For instance, for a parallel pipe system, assign arbitrary positive values to flows in the clockwise direction, and negative values to flows in the counterclockwise direction, solving for head losses, and then adding them algebraically (one is positive, one is negative). When this yields a result of zero, head losses are equal, and the correct flow has been chosen. However, this method involves many trial runs to determine the actual flows and is less direct than the equivalent pipe method.

Manual solution of real network problems can be a large undertaking. The use of the Hazen-Williams alignment nomograph can speed up the calculations, and for systems involving more than a few pipes, computer systems which automate the process are in widespread use.

Steps for solution have been outlined in the chapter. Rather than attempting more detailed explanations here, try working through some of the problems.

PROBLEMS

1. Water is flowing at 1 cfs from tank A to tank B. Tank A is 20 ft. deep. How deep is tank B?

2. One thousand feet of 8 inch diameter pipe is extended by 1200 ft. of 10 inch diameter pipe. The velocity through the 10 inch pipe is 2 ft./sec. If C=100 for both pipes, what is the head loss in this system (ft.)?

3. A cast iron pipeline, C=100, connects points A,B,C,D, which have respective elevations of 347 ft., 357 ft., 361 ft., and 351 ft. Pipe diameters and lengths are:

A to B	6 inches, 450 ft.
B to C	8 inches, 320 ft.
C to D	10 inches, 280 ft.

    Flow is .5 cfs and pressure at A is 73 psi. Determine the pressure at Point E, located on pipe BC, 100 ft. downstream from B.

4. A 10 inch diameter pipe 1400 ft. long is connected to an 8 inch diameter pipe 1200 ft. long, and then to a 6 inch diameter pipe, 1500 ft. long, all in series (C=100 all pipes).
    A. Determine the length of an equivalent 14 inch pipe.
    B. Determine the length of an equivalent 6 inch pipe.

5. A 12 inch diameter cast iron pipe ten years old carries a flow of 1 MGD for 6000 ft. After a main break, half of its length was replaced by new 12 inch ductile iron pipe. What is the head loss through the entire length of pipe at this flow (ft.)?

6. A 1000 ft. long, 8 inch diameter PVC pipe carrying a flow of 2.3 cfs reduces to a 1000 ft. long, 4 inch diameter PVC section. If the pressure at entrance to this system is 210 psi, what pressure registers on a gage at the downstream end of the system?

7. Ten miles of 20 year old CIP transmission main 18 inches in diameter carries 3.4 MGD flow toward the city. It is then extended by 2.5 miles of 18 inch diameter DIP. At this flow, what is the pressure loss (psi) across this entire system?

8. A flow of 685 gpm passes through 3 unlined steel pipes, laid in series: 3000 ft. of 14 inch diameter; 7500 ft. of 10 inch diameter; 4080 ft. of 6 inch diameter. What is the Head Loss (ft.) across this entire system?

9. A pump discharges 2 cfs into a 10 inch diameter DIP line 1600 ft. long. Just downstream from the pump discharge, a 20 ft. long, 3 inch diameter DIP is installed to recirculate part of the discharge flow back to the pump suction source. Calculate the recirculating flow (cfs).

10. An industrial pump providing 300 ft. of pressure discharges 29 cfs through 200 ft. of 18 inch diameter asbestos cement pipe; the pipe then splits into two equal length 10 inch diameter sections, each leading to an open process tank. What is the Head Loss (ft.) across one of these 10 inch diameter sections?

11. Given the system to the right:
    A. With the valve closed, what is the head loss in pipe B?
    B. With the valve open, what is the head loss in B?
    C. With the valve open, what is the head loss across the entire system?

12. Two pipes arranged in parallel admit a total flow of 3 cfs. One pipe has a 12 inch diameter and is 1500 ft. long. The other has a 6 inch diameter and is 2000 ft. long. What is the head loss across this system (C=100 all pipes)?

13. Find the water velocity in the 6 inch diameter pipe.

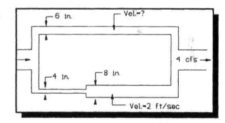

14. Three thousand feet of 6 inch diameter pipe is connected in series to 4000 ft. of pipe whose diameter is not known. If the flow through the system is .3 cfs and the pressure drop across the system is 20 psi, what is the unknown diameter (C=80 all pipes)?

15. A reservoir at 700 ft. is to supply a second reservoir at 460 ft. The reservoirs are connected by 1300 ft. of 24 inch diameter new masonry conduit and 2000 ft. of 20 inch diameter new masonry conduit, in series. What will be the discharge delivered from the upper reservoir to the lower one?

16. Which of these systems has the greater capacity (C=120 all pipes)?

    A. <u>9000 ft. (16 inches)</u>   <u>6000 ft. (12 inches)</u>   <u>3000 ft. (10 inches)</u>

    B. <u>11,000 ft. (18 inches)</u>   <u>5000 ft. (8 inches)</u>   <u>2500 ft. (10 inches)</u>

            | |
            |_____ <u>7000 ft. (10 inches)</u> _____|

17. Given the system to the right, what is the difference in elevation of the water surfaces in the two piezometers?

18. Three pipes are arranged in parallel:

Top section:	4 inch diameter, C=140, 570 ft. long
Mid section:	6 inch diameter, C=80, 900 ft. long.
Bottom section:	8 inch diameter, C=100, 1200 ft. long

The approaching flow is 3000 gpm.
A. Compute the equivalent length of a 12 inch diameter pipe with a C=100
B. What is the head loss through this system?

19. Compound pipe ABCD consists of 10,000 ft. of 20 inch, 8000 ft. of 16 inch, and L ft. of 12 inch. all in series. (C=120 all three).
A. What length of L will make the ABCD pipes equivalent to a 15 inch pipe, 16500 ft. long (C=100).
B. If the length of the 12 inch pipe were 3000 ft., what flow would occur for a lost head of 135 ft. from A to D?

20. Water entering a customer's house encounters an 80 ft. long, 3/4 inch diameter copper branch which leads to the shower. At that branch, the pressure is 40 psi. Somewhat downstream, the pipe splits into two branches, each 3/4 inch diameter copper, one 50 ft. long leading to the dishwasher, and one 30 ft. long leading to the sink. At that split the pressure is 38 psi. If shower, dishwasher and sink are all being used at once, what is the cfs flow entering the house?

21. Calculate the flow through the system to the right:

22. The flow through the entire parallel pipe system diagrammed below is horizontal, and is 6.18 cfs. What is the flow through section CD?

CHAPTER 8

MINOR HEAD LOSS

In Chapter 6 it was stated briefly that friction head loss can be separated into two types: Major Head Loss (from friction encountered along the length of the pipe) and Minor Head Loss (pressure loss at bends, valves, fittings, and changes in pipe diameter). Extra turbulence is created at these areas, and the resultant head loss is often significant enough to be included in calculations. It is almost always less than major head loss, however, and its inclusion is optional. For example, if an 8 inch diameter pipe carries water for two miles, and has only one 90 degree elbow at mid-length, the minor loss from this is considered too small to count. But inside a factory, where the piping turns, enlarges, contracts and encounters valves every few feet, the minor loss would be significant. A city water distribution system has valves and laterals at frequent intervals, and an approximation of minor loss would have to be made to determine its significance. A general guideline which can be used: if the minor head loss is less than five percent of the total head loss, ignore it.

equivalent pipe technique

Determination of minor head loss is simple and straightforward, and a standard nomograph in common use is included at the end of the chapter. For calculations, each fitting is converted to a length of straight pipe of the same diameter which would cause the same head loss. This makes further calculation easy. Once the fitting has merely been converted to an extra length of pipe, calculations for the total length will include both major and minor losses.

Inspecting the nomograph, pipe size is given as both Inside Diameter and Nominal Diameter. Although it is more appropriate to calculate using inside diameter, for practical use with a ruler here, the difference is insignificant.

Note the difference in equivalent pipe length between a fully open gate valve, and a 3/4 closed gate valve. In pumping operations, it is common practice to control flow rate by partially closing a valve on the pump discharge, thereby increasing the minor head loss, and restricting the flow. Using another example, a pressure regulating valve senses a pressure on the downstream side of the valve which is higher than its setpoint and closes slightly to increase the minor head loss, thus dropping the pressure. In water distribution systems, the importance of proper valve maintenance and record keeping is obvious. A shutoff valve in a large main left half closed by accident would significantly affect the line pressure.

Larger diameter fittings yield larger equivalent lengths of straight pipe. This seems to contradict previous calculations for friction head loss (small pipes generate more head loss), but the attempt with this nomograph is to keep the head loss value for a particular fitting the same no matter what the size. After converting that fitting to an equivalent length of pipe, head loss will then be calculated based on the diameter, and that same type of fitting would end up yielding a calculated smaller head

Figure 8.1 - Recirculated sludge lines at a wastewater treatment plant.

loss if it were on a larger pipe, unless the nomograph had designated larger equivalent pipe length for larger pipes.

The most frequent application of a Borda Entrance is at the intake end of a pump drawing a suction lift.

Borda Entrance

Note: "Entrance" refers to an entrance <u>to the pipe</u>. Similarly, "Exit" would be an exit <u>from the pipe</u>.

For enlargements and contractions, measure with the smaller pipe size on the nomograph. Two separate hydraulic phenomena are occurring in these areas of pipe enlargements or contractions:

1. Pressure drops because of extra turbulence encountered in passage to a pipe of different diameter (minor head loss). This is a permanent loss of pressure, which must be made up by the source.

2. Pressure and velocity change is encountered because the pipe changes diameter. This could be only temporary. Original pressure and velocity would return if downstream the pipe returned to the original diameter, and the only pressure drop remaining would be from major and minor head losses through this section.

Figure 8.2 - Series of tee joints manifold pumped flow.

Note: When faced with a fitting not listed on the nomograph - estimate. Use the closest one. Use care with reading nomographs; misplacing the ruler by a fraction of an inch may change the reading by a great deal.

---

velocity head technique

There are a few minor losses which may be expressed in terms of Velocity Head ($V^2/2g$). These have been worked out by experimentation and measurement, and are useful in calculations where Bernoulli's Theorem is in use and a Velocity Head value is readily at hand. This method <u>directly</u> converts to a head loss value for the fixture:

Figure 8.3 - Reduced pressure back flow preventer; a good example of Head Loss due to fittings.

Ordinary Exit:

Head Loss = 1 Velocity Head

The water loses its energy due to velocity upon entrance to the tank, because it loses all its velocity.

Ordinary Entrance:

Head Loss = ½ Velocity Head

Turbulence is created in getting the water out and around the 90 degree angle and into the pipe.

Borda Entrance:

Head Loss = 1 Velocity Head

Extra turbulence is created in getting the water out and around the 180 degree angle and into the pipe.

> **Note:** To equate the two techniques, Ordinary Exit is entered into the Equivalent length nomograph at the same place as Borda Entrance.

Most often, conversion to equivalent pipe length is the method of choice. Calculations from previous chapters remain the same when minor head loss is included. The only difference is that with this inclusion, the pipe length has been extended.

Reprinted with permission of Crane Co.

Resistance to Flow from Valves and Fittings

PROBLEMS

1. Find the equivalent length of pipe for each of the following:
   A. 45 degree elbow - 10 inch diameter
   B. 4 inch diameter Borda Entrance
   C. Sudden enlargement: 4 inches to 8 inches
   D. Ordinary Exit: 36 inches
   E. 1 inch diameter close return bend

2. What equivalent length of 8 inch diameter pipe is created at the suction pipe entrance of a well pump?

3. Water exiting a tank flows at 1.3 cfs through a 10 inch diameter galvanized iron pipe. What is the head loss (ft.) generated by the fitting?

4. The pressure drop across a 350 ft. length of horizontal 20 inch diameter DIP main is 6.8 psi. A check valve is installed on the line, how does this valve installation affect the flow?

5. Fifty ft. of 3 inch diameter pipe is connected in series to 50 ft. of 6 inch diameter pipe. If the flow is .8 cfs and is horizontal (C=100), what is the psi pressure drop across this system?

6. A 6 inch diameter main (C=110) is carrying a flow of 1 cfs. What head loss is produced by two regular flanged 45 degree elbows and an open globe valve?

7. Two hundred feet of 12 inch diameter pipe with water flowing has a slope of .02 ft./ft. There is a 3/4 closed gate valve on the pipe.
   A. What is the head loss in the pipe?
   B. What would be the head loss if the valve were fully open?

8. If all pipe is 6 inch diameter, 10 year old cast iron, and Q=3 cfs, find the total head loss through this system.

9. A .5 cfs gravity flow passes from a reservoir through 7.5 miles of 8 inch diameter 10 year old CIP to a treatment plant. About midway along the pipe is a half closed gate valve. If the incline of the pipe is 200 ft. from reservoir to treatment plant, what percent of the Head Loss is caused by the valve?

10. Water entering a house through a 1 inch diameter horizontal copper pipe makes two 90 degree turns before the pipe ends at a faucet. If the pressure at entrance is 30 psi, and the pipe is 25 ft. long, when the faucet is fully open, what is the flow?

11. Water travels from point 1 to point 2 through 1000 ft. of 10 yr. old cast iron pipe, 12 inches in diameter. Installed in the line are one swing check valve, one open globe valve, and two square elbows. If the flow is steady at 2 cfs, how much pressure does the water lose in traveling from point 1 to point 2 (psi)?

12. A tank discharges water through a 10 inch diameter pipe, 900 ft. long (C=120), at a velocity of 5 ft./sec. What is the head loss through the system:
    A. by the Velocity Head technique
    B. by the Equivalent Pipe technique

13. A 24 inch diameter transmission main carries 7 cfs for a length of two miles. Along the pipe are three 45 degree elbows and 4 open gate valves. What percentage of the total head loss is caused by the fittings on this pipe?

14. Diagrammed to the right is a heat exchanger. What is the flow (cfs) through this unit? (All pipe is 4 inch diameter copper pipe).

15. What is the elevation of the water surface in tank #2? All valves are fully open; C=110; all pipe is 12 inch diameter; Q=800 gpm.

16. A pump drawing a 20 ft. suction lift delivers 200 gpm. What is the pressure drop in the 6 inch diameter pump suction line, which is 10 year old CIP, and includes a standard elbow and a foot valve?

17. Determine the head loss in the system. All bends are standard elbows. Q=3.2 cfs.

18. Water service was extended in a galvanized iron loop to supply a temporary establishment. During pipeline construction, a shutoff valve on the line was inadvertently left half closed, as shown in the diagram. What is the flow through the loop?

19. A pump will lift 400 gpm from a river to a holding tank; The pump suction lift is 4 ft.; suction piping is 10 ft. of 6 inch diameter 20 yr. old CIP and includes two standard 90 degree els and a check valve. The discharge piping is 600 ft. of 6 inch diameter DIP. It includes one 90 degree el and a gate valve. The water level in the holding tank is 175 ft. above the river level. What pumping pressure (psi) is required to do this job?

20. An 8 inch diameter pipe (C=100) enlarges to a 12 inch diameter pipe. If the 8 inch pipe is 1200 ft. long, the 12 inch pipe is 800 ft. long (C=140), and the flow is 2 cfs:
    A. What is the head loss in the system?
    B. What is the velocity in the 12 inch diameter pipe?

21. An 8 inch diameter pipe 1000 ft. long takes water from an upper reservoir. On this pipe are a gate valve, 3 standard elbows, and a globe valve. The pipe is connected to a 6 inch diameter pipe 50 ft. long with an angle valve and a standard tee which delivers the water to a lower reservoir. If the difference in the water levels of the two reservoirs is 50 ft., what is the flow through this system (C=100 all pipes)?

CHAPTER 9

OPEN CHANNEL FLOW

Rivers, streams, aqueducts, drainage and irrigation ditches, and sewer pipes are usually only partially full. Open Channel Flow exists in conduits where the water has a free surface exposed to atmospheric pressure. Water depth being constant from one point to another, there is no pressure to act as the driving force. The only natural energy this water possesses is its Velocity Head, and at normal water velocities, this is a small value $(V^2/2g)$. The basic hydraulic principles apply, however. Open channels have a degree of roughness, just as water pipes under pressure do, and friction occurs as the moving water rubs against the sides of the channel. This quickly depletes the Velocity Head, and decreases the velocity. Therefore, open channel systems (aqueducts, sewers, drains) are constructed with a slope; the pipe is physically slanted downward just enough to overcome friction losses, and keep the velocity (and flow) constant. As long as the slope of the channel (ft./ft.) is exactly the same as the friction loss (ft./ft.), velocity will be maintained. Man here is making use of gravity to overcome friction, so that the Velocity Head can keep the water moving.

In a properly sloped channel, at steady state flow, the Hydraulic Grade Line is parallel to the water surface, and to the channel bottom. Energy loss has been negated by the slope of the channel, and the depth of the water remains the same. There is no energy change - no input due to depth (Pressure Head).

Pressure Head and Velocity Head are the same at Point 1 and Point 2, and the slope of the channel (the open channel equivalent of Elevation Head) is overcoming the friction.

If the channel were horizontal, however, the water would decrease in velocity as it encounters friction, and "pile up" behind, making its own slope. The water produces for itself the Pressure Head needed to keep moving, as a greater depth at the upstream end, and the slope of the water surface will register its progressive loss downstream.
The Hydraulic Grade Line is still drawn parallel to, and along with, the water surface,

and it registers the loss of energy. The Energy Grade Line remains horizontal: flow is steady state; as cross sectional area decreases, velocity increases.

If this channel were then slanted downward at exactly the same slope as the water surface registers, the surface will flatten out and become parallel with the conduit bottom again (as in the first diagram).

If the channel is sloped more than is necessary to overcome friction, velocity will increase, and water depth will decrease. The extra energy created by the steep slope becomes velocity energy. Enough velocity occurs to make the friction head loss equal to the elevation drop across the length of pipe ($Z_1 - Z_2 = HL_f$). As long as the flow approaching the system remains steady, the depth will decrease substantially to accommodate the new velocity.

If, in an existing channel, the flow increases, the area of flow will increase, thus holding the velocity constant. Note that in each of these situations, there is a natural adjustment of the water depth, resulting in a change in the cross sectional area of flow.

In wastewater collection systems, where the desired slope between manholes is not sufficient to maintain scouring velocity, additional slope must be provided. On the other hand, if the slope of the pipe is too steep, velocity must be reduced, and often a drop manhole is used - a method of wasting energy. Velocity suddenly drops to zero at the manhole, and for hydraulic measurements, the exit pipe is considered the upstream end of an entire new pipe system.

In natural open channels, the same phenomena are evidenced. A stream bed which slopes sharply, then levels off somewhat, will carry water at a shallow depth where the slope is steep, and velocity is high. Downstream, where the slope is gradual, the water will be deeper, and

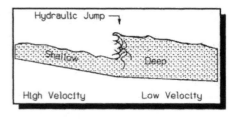

the velocity slower. Where the gradual slope was not enough to overcome friction loss, in this section the extra depth will provide enough Pressure Head at the upstream end to keep the water moving forward at a constant flow, and the depth will gradually decrease farther downstream as energy is dissipated.

In places where waters of two different velocities meet, a short section of deeper water occurs; the water level rises at the point where the velocity changes, and a "hump" forms here, before the surface smoothes out again. There is extra turbulence at the point where the two velocities meet, and the water traps air and expands somewhat. Called Hydraulic Jump, this phenomenon occurs dramatically at dams and flumes, and much less dramatically where velocity change is slight. On a small scale it is visible as ripples that occur when a stone is thrown into water, or when water pours from a faucet into a full sink or bucket.

A distinction in flow characteristics occurs between open channel flow and closed pipe flow. In a pressurized pipe, the upstream pressure determines the velocity with which the water will travel, and that velocity determines the head loss. The amount of water admitted to the system (the flow) is dependent upon this velocity and pipe diameter. In open channels, the head loss is equal to the elevation change produced by the slope of the pipe; this will determine the velocity and the depth of water in the pipe, and this velocity and depth will vary, depending upon the adequacy of the pipe slope to overcome friction losses. The total amount of flow approaching the system is admitted, however and the cross sectional area of flow along with the water velocity will adjust to accommodate this flow (unless the channel isn't deep enough, and then the water will flood out over both sides).

calculations for open channel flow

In open channels, maintenance of proper velocity is of prime importance. In order to keep the velocity constant, the channel slope must be adequate to overcome friction losses. Calculation for head loss at a given flow is necessary, and the Hazen-Williams formula is useful ($Q = .435 \ C \ d^{2.63} \ s^{.54}$). The concept of slope has not changed. The only difference is that now we are measuring, or calculating for, the physical slope of a channel (ft./ft.), equivalent to head loss.

The problem that arises is with the diameter. In conduits which are not circular (rivers, grit chambers), or in pipes only partially full (sewers, drains) where the cross sectional area of the water is not circular, there is no diameter. Another parameter must be used to designate the size of the cross section, and the amount of it that contacts the sides of the conduit. For these non-circular areas, we use Hydraulic Radius (R). It is a measure of the efficiency with which the conduit can transmit water. Its value depends on pipe size, and amount of fullness. Hydraulic Radius is determined by dividing the cross sectional area of the water, by the perimeter of conduit in contact with that water.

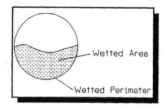

$$R = \frac{\text{wetted area}}{\text{wetted perimeter}}$$

The attempt is to measure how much of the water is in contact with the sides of the channel - or - how much of the water is not in contact with the sides, the "hydraulic" radius. The original version of the Hazen-Williams formula did incorporate Hydraulic

Radius, in fact, as did similar versions developed by Chezy, Manning, and others. The formula in use in this text, employing diameter, is a shortened version specialized for full pipe flow. For use in open channels, <u>Manning's Formula</u> has become most popular:

$$V = \frac{1.486 \; R^{.66} \times s^{.5}}{n}$$

Diameter has been replaced by R, thus making the formula flexible enough to adapt to all cross sectional areas.

Hydraulic Radius for Full Pipes:

Manning's formula can be used for full pipe flow also.

For a full pipe $\qquad R = \dfrac{\text{wetted area}}{\text{wetted perimeter}} = \dfrac{3.14 \; r^2}{3.14 \; d} = \dfrac{(d/2)^2}{d} = \dfrac{d}{4}$

> **Note:** Hydraulic Radius for a full pipe = d/4

Hydraulic Radius for Different Sized Channels

Since open channel flow includes both natural and man-made conduits, shapes can vary a great deal. Above are some typical cross sections, and below are a series of formulas to handle them in solving for R, etc.

OPEN CHANNEL SHAPES AND FORMULAS

Cross Section	Rectangle	Triangle	Trapezoid	Parabola
Area, A	wh	yh^2	$wh + yh^2$	$\dfrac{2}{3} \, hT$
Wetted Perimeter, P	$w + 2h$	$2h\sqrt{y^2 + 1}$	$w + 2h\sqrt{y^2 + 1}$	$T + \dfrac{8h^2}{3T}$
Hydraulic Radius, R	$\dfrac{wh}{w + 2h}$	$\dfrac{yh}{2\sqrt{y^2 + 1}}$	$\dfrac{wh + yh^2}{w + 2h\sqrt{y^2 + 1}}$	$\dfrac{2hT^2}{3T^2 + 8h^2}$
Top Width, T	w	$2yh$	$w + 2yh$	$\dfrac{3A}{2h}$

T = Top width of water surface
h = Depth of water

w = Bottom width of channel
y = Deals with the slope of the side of the channel in triangles and trapezoids. The side slopes y feet out for every one foot of depth.

These formulas eliminate the need for trigonometry, and can be referred to whenever needed.

Hydraulic Radius for Partially Full Pipes

The circular conduit is certainly the most frequently used open channel. Sewers are almost all circular - but they are usually only partially full of water. This leaves an odd cross section for which to find Hydraulic Radius. Solving for cross sectional area and wetted perimeter involves more complex mathematics which can be avoided by using a <u>Hydraulic Elements Curve</u> (see page 87). This is an indirect method of solving for wetted area, wetted perimeter, Hydraulic Radius, flow or velocity, for any size pipe, at any amount of fullness. <u>It does not yield these elements directly</u>. For instance, in looking up R for the partially full pipe, the curve yields the percent of a full pipe R, that the desired R is. To use the curve, measure the depth of the water in the pipe, then divide it by the full depth (pipe diameter) to obtain the Percent Depth of Flow.

Inspecting this curve:

- When a pipe is 50% full, the values of area, wetted perimeter, and flow are half of that of a full pipe. The Hydraulic Radius and velocity, however, are the same as that of a full pipe.
- When a pipe is between 50% and 100% full, the velocity and Hydraulic Radius are greater than they are in a full pipe.

Note: Hydraulic Elements Curve yields only a percent of the desired value for a full pipe.

Roughness

$$\text{Manning's formula:} \quad V = \frac{1.486}{n} R^{.66} s^{.5}$$

There is a new component - n. This is a roughness coefficient (similar to C). Materials of which open channels are constructed can be very different from those used in closed pipe flow; nature has made many of them. Man-made drainage channels are often lined with concrete, asphalt, rip-rap, or grass to prevent erosion from the flowing water. Drainage through the sides and bottom is often allowed, to prevent cracking from frost heaves or hydrostatic pressure (this quantity of seepage is very minimal when compared with the quantity of flow that the conduit is carrying, however, and will not interfere with the necessity to consider a steady state flow system). Grass lined channels are used in parks and golf courses, where aesthetics are important, or where large flows are infrequent. Curb and gutter channels are designed into paved streets, and are one-sided. They must provide scouring velocities, and have enough depth so that the water surface does not interfere with traffic. Sewer pipes can be lined or unlined, and are constructed of iron, steel, concrete, brick, stones, clay and many other materials. Manning's n factors developed as a method to approximate roughness in open channels.

> **Note:** n values decrease as the channels get smoother.

Velocity

Manning's formula equates to velocity, instead of flow. Perhaps this originated from the importance of providing scouring velocity in conduits designed to carry solids-laden water, and from the need to determine a slope that would provide that velocity. Solids allowed to deposit along the length of the channel will change the velocity, and eventually block the channel. If desired, it is easy to replace V with Q by using the Equation of Continuity $(Q=AV, V=Q/A)$.

Flow

There is a most important feature of open channel flow which must be kept in mind, and which makes it quite different in practice from closed pipe flow. For calculations the flow is always considered as steady, but the actual discharge into these channels is almost never constant, and varies over the course of a day, week, and with the seasons. Most sanitary sewers are designed so that they are half full at average daily flow. Storm drainage channels are not in use at all most of the time, yet must have extremely high capacity during wet weather. This makes design quite a challenge, and operation is not often at design efficiency.

If desired, to save time and eliminate some calculations in dealing with Manning's formula, a nomograph is provided.

MANNING ROUGHNESS COEFFICIENT, n

Type of Conduit	n
Pipe:	
Cast Iron, coated	0.012 - 0.014
Cast Iron, uncoated	0.013 - 0.015
Wrought Iron, galvanized	0.015 - 0.017
Wrought Iron, black	0.012 - 0.015
Steel, riveted and spiral	0.015 - 0.017
Corrugated	0.021 - 0.026
Wood Stave	0.012 - 0.013
Cement Surface	0.010 - 0.013
Concrete	0.012 - 0.017
Vitrified	0.013 - 0.015
Clay, drainage tile	0.012 - 0.014
Lined Channels:	
Metal, smooth semicircular	0.011 - 0.015
Metal, corrugated	0.023 - 0.025
Wood, planed	0.010 - 0.015
Wood, unplaned	0.011 - 0.015
Cement Lined	0.010 - 0.013
Concrete	0.014 - 0.016
Cement Rubble	0.017 - 0.030
Grass	----- - 0.020
Unlined Channels:	
Earth: straight and uniform	0.017 - 0.025
dredged	0.025 - 0.033
winding	0.023 - 0.030
stony	0.025 - 0.040
Rock: smooth and uniform	0.025 - 0.035
jagged and irregular	0.035 - 0.045

Manning's Nomograph

Directions for Using Hydraulic Elements Curve

- Obtain percent depth of flow (h/d).
- Follow across to the desired value, then down to Percent of Value for Full Pipe.
- Solve for this value in a full pipe.
- Multiply by the percent of value taken from the curve.

PROBLEMS

1. Water flows at a depth of 6.25 ft. in a rectangular canal 8 ft. wide. The average water velocity is 14.9 ft./sec. What is the slope on which the canal is laid if n=.01?

2. Determine the drop in elevation from beginning to end of a rectangular wooden flume 1000 ft. long, which carries 600 cfs at a 7 ft. depth. The flume measures 15 ft. across.

3. A rectangular channel whose width is three times its depth is lined with smooth concrete. When carrying a flow of 900 cfs, the average velocity is 3 ft./sec. What is the incline of this channel?

4. A 12 inch diameter lined cast iron sewer flowing full is laid on a grade of .004 ft./ft. Calculate the flow:
   A. using Manning's formula
   B. using Hazen-Williams formula

5. A reservoir feeds an open rectangular channel 15 ft. wide (n=.015). The depth of water in the reservoir is 6.22 ft. above the channel bottom. The channel is 800 ft. long, and drops .72 ft. in this length. The depth behind a weir at the discharge end of the channel is 4.12 ft. What is the flow in the channel?

6. A 48 inch diameter circular steel drain is laid on a slope of .003, and is flowing full. What is the flow?

7. The cross sectional measurements of a flume are 10 inches by 10 inches. The water is 8 inches deep, and moving at a velocity of 2 ft./sec. How many gallons of water will the flume deliver in four hours?

8. Water flows 3 ft. deep in a rectangular channel 20 ft. wide (n = .013, s = .0144). How deep would the same quantity flow on a slope of .00144?

9. Water flows at a depth of 3.9 ft. in a rectangular canal 12.5 ft. wide. The average velocity is 4.2 ft./sec. If C=100, what is the slope on which the canal is laid?

10. Which of these canal structures will carry the greater flow if both are laid on the same slope?

11. What is:
    A. the Hydraulic Radius
    B. the cross sectional area for a flow of 6 inches depth in a 10 inch diameter circular pipe?

12. A 10 inch diameter pipe is flowing 1/3 full. What is the velocity of the water if the volume of flow is 135 gpm?

13. What is the water depth in a 24 inch diameter sewer which flows at 500 gpm and 2 ft./sec. velocity?

14. Elevation of manhole #1 is 300 ft. Manhole #2, 1500 ft. downstream, lies at 295 ft. If the 18 inch diameter concrete sewer line is flowing at 10 inches of water depth, what is the water velocity in this pipe?

15. A 24 inch diameter concrete pipe on a 1/800 slope flows half full. What is the water velocity?

16. The distance from city to the wastewater treatment plant is 3.5 miles.
    A. What size concrete line (no more than 80% full) is necessary to accommodate the 2 MGD flow?
    B. How many feet will the pipe incline from city to treatment plant?

17. A 6 ft. deep flow is carried by a triangular channel with a side slope of 4 ft./ft. depth. If n=.025 and s=.006, at what velocity is the water flowing?

18. A 10 inch diameter sewer pipe with a 3 inch water depth is laid on a grade of two feet per hundred feet of length (n=.025). What is the flow?

19. A 48 inch diameter concrete pipe carries a flow of 43 cfs on a slope of 1.8 ft./1000 ft. How deep is the water in the pipe?

20. A. Determine the required diameter for a proposed sanitary sewer which must carry domestic sewage at 2.5 cfs. This maximum flow is estimated to be 250% of average daily flow. Pipe to be used is PVC (n=.009).
    B. What will be the depth of flow at average and maximum flow conditions?
    C. What is the water velocity at both flows?

21. An obstruction in a grass lined channel 6 ft. wide causes an upstream depth buildup of 2 ft. The channel bottom is horizontal and the flow is 10 MGD. How deep is the water 2000 ft. upstream?

22. A clay drainpipe flowing full slopes downward 3 ft. over a length of 1500 ft. If the water velocity in the pipe is 5 ft./sec., what is the diameter of the pipe?

23. Water from a filter leak collects on the floor beneath, then flows to a 6 ft. diameter sump from which it is carried by 150 ft. of 12 inch diameter pipe to a drainage area (n = .012 for the pipe). Free flow exists at the pipe outlet. The floor elevation is 112.0, invert of the pipe at entrance is 106.0, and outlet is 103.0. Bottom of sump is 104.0. The water level in the sump stabilized at one foot below floor level. How many gpm are being lost from the filter leak?

24. A proposed sanitary sewer must carry a flow of 125 cfs. Concrete pipe will be used. The street has a general slope of .02%, which will be followed. The minimum allowable velocity will be 1.5 ft./sec.
    A. Determine the required pipe diameter
    B. Determine the depth of water in the pipe

25. A 100 MGD water supply will be carried by gravity flow aqueduct from a mountain reservoir 150 miles to the city for treatment. Design the transport system.

CHAPTER 10

FLOW MEASUREMENT I
FLOW RATE METERS

Why measure flow? The most vital activities in the operation of water and wastewater treatment plants are dependent on a knowledge of how much water is being processed.

The municipal water works must be able to measure water made and water used in order to calculate percent lost water. Knowledge of flow is needed to size storage tanks, provide firefighting capacity, to guard against backflow, to predict future usage, and to manage water conservation programs.

Wastewater treatment plants need a flow measurement to determine solids and BOD loading to the plant, and to the receiving stream for NPDES compliance. Control of recycle rates, calculation of F/M ratio, and prediction of physical and biological unit performance is based on flow. Knowledge of incoming water quantity is vital to anticipation of plant reaction to industrial input, shock loads, and stormwater. The sizing of stormwater retention basins to prevent flooding and backup is based on a flow reading. The smooth operation of an Industrial Pretreatment Program depends on knowing the quantity of industrial input and how it compares to the quantity of the municipal stream.

Both water and wastewater utilities need flow measurements to calculate chemical feed rates, hydraulic loading to treatment units, tank detention times. It is necessary for scheduling of system maintenance, for calculation of process expenses (power, chemicals, labor), for sizing of pipes, pumps and meters. Flow records are needed to plan treatment plant expansions, to be prepared for legal proceedings, and of course, for customer billing. Flow measurement can be based on flow rate (Chapter 10), or flow amount (Chapter 11):

Flow Rate (gpm, MGD, cfs): Utilities need this type of meter inside the treatment plant, in wastewater collection, and in potable water distribution, to determine process variables. In common use are pressure differential meters, magnetic meters, and ultrasonic meters. Some are designed for metering flow in closed pipe systems; some are meant for open channel flow.

Flow Amount (gal., cu.ft.): The totalizer sums up the gallons or cubic feet which pass through the meter. Private, commercial and industrial customer service meters operate this way, and total amount of flow is the basis for billing. In wastewater treatment plants, automatic composite sampling units are usually flow proportioned to grab a sample every so many gallons. These totalizer meters may be turbine, propeller, compound or positive displacement.

Flow rate meters will be considered in this chapter, and totalizers in Chapter 11.

pressure differential meters: closed pipe flow
Orifice Meters

Orifice meters have been in use since ancient Roman times in water pipes. The principle of operation is that a pressure differential is created across the unit as a steady flow of water passes through a constricted section. As water flows from an area of large cross

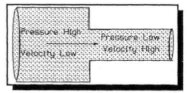

section to an area of small cross section, the velocity increases (Q=AV). Since this results in an increase in Velocity Head, and total energy must be maintained, some form of energy must decrease by the same amount; it is the Pressure Head that drops abruptly at this point. This is applied in practice to measure flow. The sensing unit of the orifice meter is a flat metal plate, usually stainless steel, installed between flanges in a pipeline,

Orifice Plate

with a precisely sized small diameter hole at the center. Pressure gages are installed one pipe diameter upstream, (ahead of turbulence at the plate), and one half pipe diameter downstream (where the jet is narrowest and pressure is minimal) to measure the pressure differential. The formula which calculates flow for an orifice meter: For derivation see Appendix IV.

$$Q = C_d \, A \, \sqrt{(Ph_1 - Ph_2) \times 2g}$$

Friction loss is incorporated as a Coefficient of Discharge (C_d), which takes in the loss of head encountered from passing through the meter (turbulence at the constriction). Its value depends upon the type of orifice chosen. Diagrammed here are several types, with their coefficients of discharge:

This formula is easy to use, but the solution can also be arrived at by using Bernoulli's Formula, and the Equivalent Pipe for Fittings chart to calculate the head loss.

Sharp Edged $C_d = .6$

Short Tube $C_d = .8$

Round Edged $C_d = .6$

Square shaped Orifice $C_d = .6$

> **Note:** the pressure drop resulting from the velocity change at the constriction is recoverable as the water continues on the discharge side of the orifice, but the head lost from turbulence created in getting through the constricted section is not.

Other useful applications of this flow formula occur wherever there is an opening in a tank or pipeline.

The diameter of the orifice is always used, and the formula applies even in cases where water flows freely from the structure to the atmosphere. In this situation, $Ph_2 = 0$. Ph_1 in the tank or pipe, is the pressure at that point.

Orifice meters are inexpensive and easy to install. They require little space, and are widely used in industry for water, steam and gas flow measurements. However, there are some important disadvantages:

Orifice meters are not extremely accurate. The device causes a great amount of turbulence on discharging from the orifice which results in large head loss to the whole system. Note that the presence of the Coefficient of Discharge in the formula could decrease the flow by 40%. In addition, the greater the difference between the two diameters, the greater will be this turbulence, and resultant head loss. This is not accounted for in the formula. Small initial cost must be weighed against cost of pumping, and if high pressure and specific flow quantity are important to the system, this may not be the metering device of choice.

Corrosion and solids in the water can shortly wear a sharp edged orifice into a round edged orifice, substantially changing the C_d value (C_d goes up, but by how much?).

Orifice meters are not recommended for permanent installation to measure wastewater flows; solids in the water easily catch on the orifice, throwing off accuracy.

For installation, it is necessary to have ten diameters of straight pipe ahead of the meter to create a smooth flow pattern, and five diameters of straight pipe on the discharge side. In some installations where a meter is desired, this is not possible.

Venturi Meters

Venturi meters, like orifice meters, measure flow in closed conduits by recording a pressure differential across a constriction in the pipe. The meter consists of a constricted tube, complete with pressure gages or manometer, and is purchased as a unit. It has a converging section, a throat (where fluid velocity increases), and a longer diverging section. The gages are placed just upstream from the convergence, and at the throat. The pattern of flow is smooth through the unit, and there is very little pressure loss due to turbulence. The angle of divergence is about 7%, which does require the tube to have substantial length. Like the orifice, the flow formula for venturi meters is derived from Bernoulli, and it was he, in the 1700's, who first described its operation. Calculation would be the same as for the orifice, except for one important difference. If the diameters of entrance pipe and throat are not very different, the values of the velocity heads at these two points of measurement are similar. Therefore, Vh_1 cannot be ignored. However, if the throat diameter is ½ the upstream diameter, or less, the inclusion of Vh_1 is insignificant, and a good approximate flow may be obtained using the orifice formula.

Flow formula for Venturi Meter:

$$Q = C_d \; A \; \sqrt{\frac{(Ph_1 - Ph_2) \times 2g}{1 - \left(\frac{d_2}{d_1}\right)^4}}$$

For Derivation see Appendix IV.

The venturi meter is the most accurate and the most popular closed conduit differential pressure meter. It can be used over a wide range of flows, and has a low permanent head loss because of the smooth configuration ($C_d = .98$). It is applicable to the measurement of wastewaters and sludges, for there are no protruding parts to catch solids or wear rapidly. Maintenance is a matter of flushing, and cleaning solids out of the pressure gage tubes. Calibration is done by testing at a known flow rate.

Disadvantages of the venturi are its long length, and high capital and installation costs.

Flow Rate Controller

In water filtration facilities which operate constant rate filters, the rate of flow through the filters is regulated by a valve/venturi mechanism, the Flow Rate Controller. It is installed in the filter effluent pipe, and responds to flow changes through the filter sand. As the pores in the filter media fill up with suspended matter, head loss in the bed increases, and less water flows through. At the beginning of the filter run, the controller valve is almost shut. Water passing through the filter has little resistance, and flow would be high, except that passing through this almost-closed valve it encounters enough head loss to drop the rate to a predetermined set point. As the filter bed clogs, passing less and less flow through, the controller valve responds to the dropping pressure from the venturi, and gradually opens, thus keeping the flow constant.

Operating Mechanism: Adjustment weight on calibrated lever at top is preset to desired flow, and aids valve opening (pulls up). Pressure from flow passing through body of venturi has flexible diaphragm in depressed position at beginning of run, and valve is almost closed. Lesser pressure underneath diaphragm from venturi throat contributes to positioning. A pressure differential is created above and below diaphragm, which determines its exact position. As pressure decreases during the run, this differential decreases, and diaphragm is allowed to move up, slowly opening the valve, and maintaining the flow at a constant rate. Modern flow rate controllers are usually computer controlled and programmable. The advantage of automatic operation is that the set point for valve movement can be adjusted from the control center, movement of the valve is more precisely controllable, and the controller can be set to sort out and respond to a number of other inputs (turbidity, temperature, initiation of backwash, high water alarm in clearwell). In the past, a series of relays was set up to respond to multiple inputs, but fluctuating response from the many inputs (searching) would result. With the Programmable Controller, a clear decision is made, and acted upon.

Manometers

Manometers can be installed in venturi meters or orifices instead of the pair of pressure gages. A manometer is a piezometer tube curved in a U shape, and filled with a liquid of known Specific Gravity. Pressure on one end of the tube causes the liquid to move, by an amount equivalent to the pressure.

Simple Manometers are connected to the pressure source at one end, and the internal liquid is free to ride up the other end, as the pressure moves it. These simply read pressure. Manometers installed on anaerobic digester gas lines are of this type. They are also often seen installed on liquid chemical storage tanks, and are used as a means to determine liquid level. The internal fluid may be brightly colored for visibility. The barometer, a manometer which responds to air pressure, is of this type also. It is in contact with the atmosphere at one end (the pressure source), and the free end is enclosed in vacuum. The unit is placed up against a calibration scale so that the liquid movement can be measured.

Differential Manometers are connected at both ends to different sources of pressure. Installed into a venturi meter, the device would provide not just a pressure reading, but a pressure differential between the two sources. The center liquid can ride back and forth, depending on which side has the greater pressure. By attaching the differential manometer to both high and low pressure sides of the venturi constriction, the difference in pressure can be measured, and then converted to flow rate. Installed this way onto water pipes, the pressure source (water) must be kept separate from the center fluid by an impermeable membrane.

The two most commonly used center fluids in differential manometers are water and mercury. Water is used as the fluid (colored for visibility) when the pressure differential is expected to be less than 1.25 psi. If the pressure difference were 1 psi, for instance, it would move the internal fluid in a water manometer 2.31 ft. Anything much larger than this would be unwieldy.

Mercury is most often used in manometers as the center fluid. It is 13.6 times heavier than water (S.G. = 13.6) and it takes that much more force to produce the same amount of movement as in a water manometer. Thus, with a significant pressure differential, the fluid in mercury manometers only moves a matter of inches. This is much more readable.

To calibrate a manometer, start with both ends connected to exactly the same pressure (or disconnect it completely) and let the center fluid ride to the middle of the tube, so that it rises to the same level on both sides. Mark that as the zero point on the calibrated scale. Now connect it up to the two pressure sources, and read the rise or drop in fluid level. If the center liquid is mercury, inches of movement on the scale will be read. Convert it to feet. This is now <u>feet of mercury</u>. Convert it to <u>feet of water</u> (multiply by 13.6, the Specific Gravity of mercury). To solve for flow through the unit (if this is a venturi manometer), insert it into the formula for flow through venturi meters. This differential pressure reading, in feet of water, <u>is</u> (Ph_1 - Ph_2).

If the center fluid is water, read the differential, in inches. Convert it to feet of water. The Specific Gravity of water is 1, so there is no need to adjust for this. If the center fluid is another substance, multiply by the Specific Gravity of that substance.

pressure differential meters - open channel flow
Weirs
To measure flow in sewers, streams, drains - open channels - weirs have become most popular. They are merely flat or notched metal obstructions with sharp top edges, inserted across the line of flow. The water builds up behind the height of the weir, it flows over, and the depth of the overflow water is measured. The depth of water going over the weir is referred to as the "head" on the weir. Since it will be deeper at greater flows, calculations can equate it to flow, and shortened formulas for flow over weirs of specific shapes have been developed.

The two most commonly used weir types are:

Rectangular Weir - for large flows.

$$Q = 3.33 \ L \ h^{1.5}$$

h = head on weir; it is measured from edge of weir in contact with the water, up to the water surface.

90 Degree V Notch Weir - for small flows

$$Q = 2.5 \ h^{2.5}$$

h = head on weir, and is measured from bottom of notch to water surface.

From a maintenance point of view, weirs may be thought of as the open channel equivalent of orifice meters; they have some of the same characteristics. Inexpensive to purchase or make of sheet metal with a wooden backup board for strength, they are easy to install and remove, are adaptable to a wide range of flows, and are very commonly used for temporary measurements of stormflows. They can be permanently installed, but solids carried by the water will catch on the lip or V notch, and decrease measurement accuracy. Frequent maintenance must be planned where weirs are in use.

Figure 10.1 - Effluent weir showing free fall.

A depth gage (calibrated stick) is usually used to take the head measurement. In permanent installations, electronic level sensors can be used. Water flowing over a weir contracts at the point of overflow, and the head measurement must be taken upstream of that contraction. Water flowing over a V notch weir must stay within the notch for accurate measurement. If it doesn't, a larger V notch, or a rectangular weir should be used. Water flowing over any weir must fall free of the weir plate (not dribble down the side) for good accuracy. If it doesn't fall free, a weir with a smaller notch should be used.

Figure 10.2 - Clarifier exit weirs.

Weirs are often seen used in clarifiers, grit chambers, and other treatment plant units to reduce velocity or control depth in the unit. Saw toothed weirs, consisting of many small 90 degree V notches, are used as effluent structures in clarifiers.

When weirs are installed in sewers and manholes, this may cause backup problems, especially at times when stormwater enters the system. In addition, the extra depth created behind the weir produces a greater cross sectional area of flow; therefore, velocity is decreased, in effect to zero, and solids will settle and collect here, throwing off measurement accuracy.

> **Note:** Installing a weir in a channel will effect deeper water (the height of the weir plus the head on the weir) immediately upstream of the weir.

What happens downstream from the weir? Below is an exaggerated diagram of a sewer pipe with a weir installed. The slope of the channel is the same throughout.

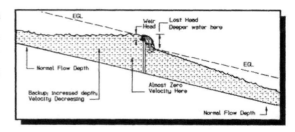

Parshall Flumes

The flume is similar to the venturi meter, except that it is meant to measure flow in open channels. The Parshall Flume is the most popularly used type, and is much used for the measurement of wastewater and irrigation water, due to its low permanent head loss, and self cleansing capacity. Its geometry is smooth flowing; solids will not get caught on it, and it operates well at both high and low flows.

The Parshall Flume has three sections:

Inlet Section - with rapidly converging walls and a level bottom. The decreasing diameter and lack of slope produces a deeper water here, and an optimal location for depth measurement for calculation.

Throat - the narrowest part, with a downward incline. Flumes are sized based on throat width (they range from 1 inch to 50 feet).

Outlet Section - diverging and sloping upward.

At low flows, water flows freely down the flume and out. Formulas for calculation are based on this condition. At higher flows, a hydraulic jump forms at the outlet section, and as the flow increases, the jump will move upstream until it completely submerges the throat, restricting the flow.

Figure 10.3 - Parshall flume at headworks of wastewater treatment plant.

A depth measurement is taken at the first 1/3 of the inlet section with a calibrated stilling well (piezometer at the side of the flume), or one with a float gage, or by overhead ultrasonic unit. The entire depth of the water is measured. If flow is free and there is no submergence of the throat, depth at the well is measured and flow is estimated:

$$Q = 4 \ W \ H_a^{1.52} \ W^{.026}$$

W = width of throat

H_a = depth in stilling well upstream

Formulas for flow through Parshall Flumes differ, dependent on throat width. The one above is used for widths of 1-8 feet, and applies to a medium range of flows.

If the downstream depth drowns out the jump and submerges the throat, the flow will be restricted, and the formula will not yield the true value. The correction depends on the percent submergence, and in flumes where this is a possibility, another stilling well is installed at the end of the throat section, so that this depth also can be measured. The percent submergence is obtained by dividing the downstream depth by the upstream depth, H_b / H_a , and the submergence graph below is referred to.

Submergence, H_b/H_a, in Percentage

Directions for Submergence Graph

• Calculate flow from formula, using H_a for depth.
• Find % submergence (H_b / H_a).
• Enter graph at bottom; move up till proper throat width is reached, then left to Q/Q_o (a percent by which the flow decreases).
• Multiply Q/Q_o by flow calculated from formula - yields corrected flow.

Note: Only when the throat is more than 50-60% submerged will a correction be needed.

Other formulas for Parshall Flumes

Throat Width (W), in.	Flume Formula
1	$Q = 0.338\ H^{1.55}$
2	$Q = 0.676\ H^{1.55}$
3	$Q = 0.992\ H^{1.547}$
6	$Q = 2.06\ H^{1.58}$
9	$Q = 3.07\ H^{1.53}$
12-96	$Q = 4\ WH^{1.522\ W8.874}$

Maintenance on a Parshall Flume is minimal. However, it is important to keep the stilling wells free of debris.

The Parshall Flume is the most widely used type of flume for measuring flow, but it is not the only kind. The Palmer-Bowlus flume has had considerable application. Its advantages are that it is smaller in size, does not require a drop in the floor of the channel, and can be prefabricated from fiberglass, plastic, steel, etc. It is useful for temporary or permanent flow measurement at manholes.

trajectory of a jet

Flowmeters are not installed in every location where one might want to measure flow; fortunately, there are some simple methods of estimating quantity. For the locations where water freely exits the end of a pipe or channel, and drops to a surface below, flow can be quickly computed based on measurement of the horizontal and vertical distance of travel after exit.

For Derivation see Appendix IV.

$$Q = 4\ A\ \frac{x}{\sqrt{y}}$$

Note: If the pipe is not full at exit, the Hydraulic Elements Curve is needed to calculate area. If the channel is shaped other than round, a formula for the area of that shape is needed.

These are specialized instances, but more common than one might think. Inplant, for recirculating flows, and for chemical feed, water often flows freely into a tank.

In the absence of flowmeters, discharge can also be estimated by filling or emptying a tank of known volume, and timing the process. For very small amounts, clock the amount of time it takes to fill a cylinder or bucket.

pitot gage (pitot tube)

Invented in the 1700's, this is a flowmeter which relies on a direct measurement of Velocity Head, based on differential gage readings, as explained in Chapter 4. The section facing upstream senses Pressure Head and Velocity Head. The downstream side only measures Pressure Head. Subtracting for Velocity Head, this can easily be converted to a flow rate.

The Pitot Tube is used to measure system and fire flow, and can be attached to fire hydrants in water distribution systems.

Fire Flow Testing

This is a practical application of pitot gage use: to determine fire flow or system head conditions during fire flow, to predict pipe roughness and need for line cleaning, and to assess risk of backflow conditions.

1. Connect pressure gage to a hydrant. This is the residual hydrant. All pressure readings are to be taken here, and it should be farther from the water source than the hydrant to be flowed, but on the same main.

2. One or more hydrants upstream are opened and flowed. Flow is measured using a pitot gage.

3. Measure pressure at residual hydrant before and during flowing. To get an accurate reading, make sure there is at least a 20 psi drop in pressure between static and flow conditions on the residual hydrant.

4. Solve for fire flow. This is one formula which can be used:

$$Q_f = Q_t \left(\frac{P_s - P_f}{P_s - P_t}\right)^{.54} \quad \text{(use gpm \& psi pressure)}$$

    Q_f = fire flow
    Q_t = discharge during test (calculated from pitot)
    P_s = residual pressure with no hydrants open
    P_t = residual pressure during test
    P_f = residual pressure during fire conditions (subtract minimum allowable pressure from P_s)

This variation works just as well:

$$\frac{Q_f \text{ (fire flow)}}{s^{.54} \text{ (under fire flow)}} = \frac{Q_t \text{ (test flow)}}{s^{.54} \text{ (under test conditions)}} \quad \text{(use cfs \& ft. head)}$$

Cross multiply and solve for fire flow.

The formulas are used in the industry to obtain accurate fire flow capacity. A ratio is established between head losses at test conditions, and at fire flow conditions to obtain the maximum flow that can be achieved without dropping the pressure below minimum allowable level (usually 20 psi). It eliminates C factor and pipe diameter, the two values which are usually not exactly known.

magnetic flow meter

Widely used in wastewater treatment plants, magmeters consist of a set of magnetic coils and an opposed pair of electrodes which are purchased as a flanged pipe unit and mounted in the process pipeline. The magnets are inserted onto crown and invert of pipe, and throw a magnetic field right through the pipe. The electrodes protrude through either side of pipe wall.

It works on the principle that when an electrical conductor (the water) passes through an electromagnetic field, a voltage is induced at right angles to the field, and this voltage is proportional to velocity. It is picked up by the electrodes, and then converted to a current signal, which can read out as a flow rate.

The magmeter offers no restriction to the flow, can be installed into short runs of pipe, is not affected by liquid properties. The electrodes must be kept free of solids. There is also an open channel version of the magnetic meter, used to measure flows in wastewater collection systems.

ultra sonic meters

More recently developed, ultra sonic meters are being used to obtain flow readings in pipe systems, and in open channels.

For closed pipe flow, the transmissive type of meter sends sonic beams through the pipe from opposite transmitter/receivers mounted at a diagonal to the flowstream. The difference between upstream and downstream directed beams is measured, is directly proportional to velocity, and is calculated to flow. This meter is installed completely outside the pipe, and the pipe wall thickness must be exactly known. In use for clean water lines only, solids in the water will interfere with the flow reading. The reflective type of sonic meter is meant for wastewater use. A single sonic beam is sent into the water from a transducer mounted on the pipe; it bounces off solids in the water, and returns at a different frequency because of the water velocity. The magnitude of frequency change is measured, velocity is computed, and then flow.

For open channel flow - an overhead unit sends ultra sonic beams down to the water surface. This is often located at the head of a Parshall Flume in a wastewater treatment plant (where the stilling well would be). The beam bounces off solids in the water and returns to the sensor. Flow is determined based on the time it takes for the beam to return. It is actually measuring the depth of the water, and then calculating for flow. Restriction is that there must be solids in the water.

rotameter

Rotameters are adaptable to small flows; they are accurate, but expensive, and are used most often on chemical feed lines. This is a variable-area meter, whose operation is based on the continuity principle ($Q=AV$). The flow passes through a transparent, calibrated, diverging tube in which is suspended a float. The velocity of the water carries the float (heavier than water) up the tube until the area is wide enough to slow the velocity to the point where the float remains suspended and stationary. At that point one looks at the calibrations to read the correct flow.

Figure 10.4 - Rotameter measures flow on a chlorinator.

PROBLEMS

1. Water passing through a 2 inch diameter new sharp edged orifice has a velocity of 22.5 ft./sec. and discharges to the atmosphere. What is the head causing flow?

2. Water exits a 4 inch diameter fire hose through a tubular nozzle with a single one half inch opening. If the pressure in the hose is 60 psi, what is the velocity of the water passing through this opening?

3. Corrosion creates a 1 inch diameter hole in the bottom of an elevated tank. The distance from ground to tank bottom is 60 feet. If C_d = .80, and 100,000 gpd is lost from this leak, what was the depth of the water in the tank when the leak started?

4. A 36 inch diameter water pipe with a pressure of 60 psi rubs against a rock and develops a 2 inch diameter hole in its side. How much water is lost in a day through this hole? (gpd)

5. A venturi meter reads 60 psi above the meter and 40 psi at the throat. Pipe diameter is 8 inches. Throat diameter is 2 inches. Calculate the flow (cfs)?

6. What should be the minimum weir height for measuring a flow of 1000 gpm with a 90 degree V notch weir, if the flow is now moving at 5 ft./sec. in a 2 ft. wide rectangular channel?

7. A weir 3 ft. high extends 20 ft. across a rectangular channel in which there is 100 cfs flowing. What is the water depth just upstream from the weir?

8. A 90 degree V notch weir is to be installed in a 24 inch diameter sewer to measure 500 gpm. What head should be expected?

9. A primary clarifier 30 ft. wide discharges .5 MGD over a weir the width of the tank. How high is the head on the weir (inches)?

10. What size orifice is required to discharge .565 cfs under a head of 28.5 ft.?

11. What is the flow through a 20 ft. by 16 ft. venturi meter with a 2 inch reading on a water differential manometer?

12. Water will exit a 20 ft. deep storage tank through a square cut hole to a flow-through process tank 4 ft. below. If the desired flow through the process is 5 cfs, what should be the dimensions of the hole?

13. A mercury manometer shows a 3 inch pressure difference across a 1 inch diameter orifice plate installed in a 6 inch line. If C_d = .68, what is the flow through this orifice?

14. A differential mercury manometer is installed on a pipe at an orifice meter. The differential reading across the orifice is 4 inches. What is the pressure difference in feet of water?

15. What is the gage difference on a mercury manometer, in inches, for a 3 cfs discharge through a 10 inch by 4 inch venturi meter?

16. A 6 inch by 2 inch venturi meter produces a 3 inch pressure differential on a mercury manometer.
    A. What is the pressure difference in feet of mercury?
    B. What is the pressure difference in feet of water?
    C. What is the flow through the venturi?
    D. What is the velocity of water at the throat?
    E. What is the velocity of water in the pipe upstream from the meter?

17. At the head works of a wastewater treatment plant, a Parshall Flume with a throat width of 3 ft. has a throat submergence of 18 inches and an upstream depth of 20 inches. What is the wastewater flow in cfs?

18. A 1 inch square hole develops in a blind flange. The pipe is 4 ft. off the ground. Pipe pressure is 80 psi. At what distance in front of the pipe will the ground get wet?

19. A 12 inch pipe flowing 1/3 full discharges from its end a stream of water which travels 2 ft. horizontally and 3 ft. vertically. How many gallons per day flow from the end of this pipe?

20. Water exiting the end of a hose at 1 cfs projects 3 ft. outward before hitting the ground 18 inches below. What is the diameter of the hose (inches)?

21. A tank 12 ft. long and 4 ft. wide contains 4 ft. of water. How many minutes will it take to lower the water to a 1 ft. depth if a 3 inch diameter plug (C_d = .6) is opened at the bottom of the tank?

22. A 48 inch diameter pipe with a flow of 2 cfs, a slope of .02 ft./ft., and a water depth of 6 inches, has a 90 degree V notch weir installed, whose height is 1.5 ft. to the bottom of the notch. For how many ft. will the water back up behind the weir?

23. An overflow masonry dam is to be constructed across a stream. Maximum flood flow is estimated to be 30,000 cfs. Six sluice gates 8 ft. high and 6 ft. wide (C_d = .85) are to be constructed in the dam (their sills at elevation 1122 ft.). The overflow weir will be 200 ft. long with a crest elevation of 1184. With all gates open under flood conditions, what will be the elevation of the water surface behind the dam?

24. A pitot tube in a pipe in which water is flowing is connected to a differential manometer containing water. If the difference in water levels in the manometer is 3.5 inches, what is the velocity of flow in the pipe ($C_d = .9$)?

25. A U tube which is open at both ends is partially filled with mercury. Water is poured into the left side of the U tube until the volume of water flows over the top. It is 2 ft. deep in the tube. Calculate the distance from the top of the U tube to the mercury in the right side of the tube.

26. Water flow for firefighting is being tested at a hydrant in a residential area. If during the test, pressure at the residual hydrant 300 ft. downstream is 60 psi static, and 43 psi under test conditions, what flow can be delivered for fire fighting at 20 psi if 1000 gpm was delivered during the test?

CHAPTER 11

FLOW MEASUREMENT II
TOTALIZER METERS

This chapter will consider the service meters; equipped with totalizers, these meters count up the flow as it passes through the unit - for estimating water demand, and for billing.

positive displacement meters

Positive displacement meters are suitable for residential customers who experience low flows and long periods when no water is used. The operating mechanism is a

Figure 11.1 - Inventory of customer meters and yokes.

measuring chamber which encloses a disc or piston. Each time a unit of water passes through the meter, the disc oscillates to pass the flow, and transmits the movement to a register, which records. The register unit is permanently sealed, and contains the register and a reducing gear train. A magnetic contact from the piston to the register unit drives it. Positive displacement meters are accurate at low flows, but have high head loss (approximately 5-10 psi), and are too expensive to use for mainline flow. They are normally used in 5/8 inch to 3 inch sizes, and normal flows should not be more than 1/3 of maximum meter capacity; operating them at excessive flows will result in wear and underregistration. Most residential services are metered with 5/8 inch meters that have 3/4 inch connections.

turbine meters

Turbine meters have a rotor in the measuring chamber which turns in response to the velocity of the water. There is either a direct linkage to the receiver, or the turbine blade may be fitted with a small magnet, and each revolution is picked up by a sensor on the outside of the meter. The movement is registered on a drive train which totalizes flow. Turbine meters are designed to handle large flows. The unit is inaccurate at low flows; small amounts of water slip past the rotor without moving it. Turbine meters generate very little head loss. The newer, more efficient models are called Turbo Meters, or Class I Turbines.

compound meters

Also a customer meter, the Compound Meter is most useful where accuracy at both high and low flows is required. Within its body are a turbine meter on the main line, and a positive displacement meter on the bypass. At low flow, water passes through the positive displacement section, and the recording is of this section. The water continues on through the turbine section, but no recording takes place. When the pressure becomes great enough, the bypass shuts off and the turbine takes over, actuating the register. A disadvantage of this type of meter is the momentary loss of reading during the changeover from low to high flow.

Compound Meters are the most expensive of the service meters, but if 10-30% of the flow is on the small side, the savings achieved in recording those low flows will probably be worth the cost. If not, a turbine should be installed.

propeller meters

Similar to the turbine meter, this unit is used in water mains where totalizing is required. Propeller meters are designed for constant high flow operation, and minimal pressure loss. The propeller is small, and mounted right in the main, facing upstream. Water velocity turns it and the register converts it to flow and totalizes.

proportional meters

To measure fire flow water, a proportional meter is used. For testing of firefighting equipment, these are installed onto fire hoses. The metering apparatus is a small turbine or positive displacement unit installed into a loop which extends off the main line, and measures a small percentage of the water flowing

through the line. A multiplying factor is built into the meter so that an accurate reading of the whole flow can be obtained. Fire line meters were designed this way because fire insurance companies require that no restrictions which could interfere with fire flow be installed onto the line (even if the meter were broken).

sizing service lines and meters

It is up to the discretion of the water utility to decide which and how many customers will be connected to the community system. Once connected, however, it is the responsibility of the utility to supply adequate pressure and flow to all customers, whether or not it is a metered system. There must be enough water to satisfy peak demand and fire flow, and pressures must be high enough throughout the system to minimize the possibility of backflow. The utility's responsibility extends up to the customer meter; although is not responsible for pressure losses within the customer's building, reasonable loss should be allowed for. When installing service to a new development or industry, the utility usually designs the service line so that there will remain in the line a pressure of at least 35 psi immediately downstream of the meter.

The type of demand must be considered. Industry usually requires a continuous flow. Irrigation water is the largest demand, but is of short duration. Domestic demand is the smallest, and use is intermittent.

1. Establish a peak flow demand (gpm needed) for the new establishment. Domestic water demand is due to the use of plumbing fixtures and washdown facilities by the customers. Assuming an average pressure of 35 psi available at the meter, the Plumbing Fixture Value Table lists various fixtures and a flow value for each.

•Take an average dwelling; refer to the table.
•Obtain Fixture Value (flow needed) for each fixture in the house.
•Multiply each by the number of fixtures of that type which will be in the house.
•Add them all up.

This is the peak flow demand (gpm). It would be valid if all residents of this dwelling use all these fixtures at once. The probability of this is slim. Therefore, make correction from Water-Flow Demand Per Fixture Value chart, and obtain a more realistic peak demand value for the dwelling. Multiply this value by the number of dwellings expected of this type.

- Do the same for other types of dwellings in the new area.
- Do it also for the industries.
- Add them all up. This is the peak demand. Use it as the flow.
- It may be advisable to adjust for projected future demand.

For an industry, often it is possible to start with the flow that this industry knows it will need, and eliminate this whole section for it.

2. Size the service line (lateral which runs from the main to the new establishment).

Information needed:
Flow capability of main that serves the area.
Static pressure in main during peak demand periods.
Elevation difference between main and establishment.
Length of service pipe from main to meter.
Friction loss through meter and valves.
Size of customer's piping and approximate head loss within building.

- Measure elevation difference from main to establishment; convert to psi and add or subtract from main pressure.
- Draw picture of the new lateral with all fittings and valves. Measure length. Choose a diameter. If there is more than one pipe size involved, include it. Draw it out exactly the way it will be installed.
- Convert all fittings and valves to equivalent lengths of straight pipe. Add up. Add to pipe length.
- Using the peak demand flow just calculated, and the Hazen-Williams formula, obtain slope, then head loss. Convert to psi.
- Subtract the head loss from the static main pressure. This is the pressure at the end of the lateral, just upstream of the meter.
- If this is an industry, does it require an RP device? Subtract the psi lost from this.

Is there still at least 35 psi pressure left?
If not, choose a larger diameter pipe for the lateral, and try again.

There is one more thing to consider: At this flow and pipe diameter, calculate the velocity. It should be under 15 ft./sec. If it isn't, choose a larger lateral to slow down the water. The necessary flow will still be obtained, the pressure drop will be less, and hammer and corrosion will be minimized.

> **Note:** These calculations are theoretical, based on minimum pipe size that will do the job. Piping material and cost of different diameter piping come into play here also, and are often a determining factor.

NEXT: - Choose a meter

Choose a meter of the type and size that is right for your flow and type of use. Meters should not be chosen by pipe size, or by pressure loss. Each meter type and size has been designed for a specific range of flows. AWWA has established standards for cold water meters based on flow capacities. The chart, Meters By Flow, lists various meters, their capacities and uses. The chart, Meter Head Loss lists average pressure losses for different meters.

A meter which is too large for the flow is a waste of money, and will let small flows slip through unrecorded.
A meter which is too small for the flow which must pass through it will work too hard and wear out early.
The meter chosen should have a maximum capacity of at least twice as much as the expected flow.
Check the listed psi loss through the meter which is chosen. Subtract this from the static pressure in the lateral which has just been designed. If there is still 35 psi left, the expected flow will be delivered to the establishment.

> **Note:** Meter size/type should be based on expected flow.

Plumbing Fixture Value

Fixture Type	Fixture Value Based on 35 psi at Meter Outlet
Bathtub	8
Bedpan washer	10
Combination sink and tray	3
Dental unit	1
Dental lavatory	2
Drinking fountain (cooler)	1
Drinking fountain (public)	2
Kitchen sink: ½-in. connection	3
3/4-in. connection	7
Lavatory: 3/8-in. connection	2
½-in. connection	4
Laundry tray: ½-in. connection	3
3/4-in. connection	7
Shower head (shower only)	4
Service sink: ½-in. connection	3
3/4-in. connection	7
Urinal: Pedestal flush valve	35
Wall or stall	12
Trough (2-ft. unit)	2
Wash sink (each set of faucets)	4
Water closet: Flush valve	35
Tank type	3
Dishwasher: ½-in. connection	4
3/4-in. connection	10
Washing machine: ½-in. connection	5
3/4-in. connection	12
1-in. connection	25
Hose connections (wash down): ½-in.	6
3/4-in.	10
Hose (50-ft. length - wash down): ½-in.	6
5/8-in.	9
3/4-in.	12

Reprinted from *Sizing Water Service Lines and Meters*, M22, by permission.
Copyright 1975, American Water Works Association.

Water Flow Demand per Fixture Value - Low Range

Water Flow Demand per Fixture Value - High Range

Meters By Flow

Displacement Meters

5/8 inch	1.0 to 20 gpm - continuous to 10 gpm (homes, small business)
3/4 inch	2.0 to 30 gpm - continuous to 15 gpm (homes, apartments, small business)
1 inch	3.0 to 50 gpm - continuous to 25 gpm (apartments, small business)
1½ inch	5.0 to 100 gpm - continuous to 50 gpm (motels, large apartments)
2 inch	8.0 to 160 gpm - continuous to 80 gpm (large hotels, housing complex)

Turbo Meters

1½ inch	4 to 160 gpm - continuous to 120 gpm (small industry, motels)
2 inch	3 to 200 gpm - continuous to 160 gpm (hotels, housing complexes)
3 inch	5 to 450 gpm - continuous to 350 gpm (hotels, industry, irrigation)
4 inch	15 to 1250 gpm - continuous to 1000 gpm (large industry, irrigation)
6 inch	30 to 2500 gpm - continuous to 2000 gpm (processing, irrigation)
8 inch	35 to 4400 gpm - continuous to 3500 gpm (processing, pump)
10 inch	55 to 7000 gpm - continuous to 5500 gpm (processing, pumps)
12 inch	400 to 4300 gpm - continuous to 2150 gpm (processing, pumps)

Compound Meters

2 inch	1.0 to 200 gpm - continuous to 160 gpm (motels, schools)
3 inch	1.5 to 450 gpm - continuous to 320 gpm (hotels, large hospitals)
4 inch	2.0 to 1250 gpm - continuous to 500 gpm (apartments, dormitories)
6 inch	2.5 to 1250 gpm - continuous to 500 gpm (apartment complexes)
8 inch	3 to 1600 gpm - continuous to 800 gpm (large condominiums)
10 inch	4 to 2300 gpm - continuous to 1200 gpm (large complexes)

Meter Head Loss

Displacement Meters

Size	Pressure Loss (psi at gpm)
5/8 inch	7.0 at 20
3/4 inch	9.0 at 30
1 inch	11.0 at 50
1½ inch	11.5 at 100
2 inch	12.0 at 160

Compound Meters

Size	Pressure Loss (psi at gpm)
2 inch	5.0 at 160
3 inch	4.0 at 320
4 inch	3.0 at 500
6 inch	14.0 at 1000
8 inch	14.5 at 1600

Turbo Meters

Size	Pressure Loss (psi at gpm)
1½ inch	15.4 at 120
2 inch	5.6 at 160
3 inch	6 at 350
4 inch	3.6 at 1000
6 inch	6 at 2000
8 inch	8.3 at 3500
10 inch	6.2 at 5500

PROBLEMS

1. Total metered community water use for a 24 hour period is 5,207,600 gallons. If the community population is 40,000, what is the gpd/capita use?

2. An industrial meter is tested and found to be underregistering by 8%. If the industry uses 200,000 gpd and pays $20.00/1000 cu.ft., how much revenue is the utility losing annually because of the faulty meter?

3. To prevent freezeup, a metered customer is allowed a let run at a rate of 2 gpm. If the customer is not to be billed for this water use, and the utility charges $.03/cubic foot registered on the meter for water, how much should be deducted from the customer's monthly bill while the let run continues?

4. The utility for an unmetered residential community wishes to account for lost water. The 50 ft. diameter cylindrical water storage tank has a drawdown of 5 ft. during the hours 12:00-3:00 a.m. The metered flow during that time period at the wastewater treatment plant is 380 gpm. What is the percent lost water?

5. A Michigan water system produces 2 MGD. If 20% of water made is lost through leaks, and each customer uses an average of 65 gallons per day, how many people does the system serve?

6. If it takes 20,000 gallons of water to produce 1000 gallons of gasoline, and the average car gets 25 miles to the gallon, and the gasoline company pays $2.00/1000 gallons of water to make the gas, how many driving miles can be obtained for a dollar's worth of water?

7. The city of Gladrock will expand its water system to service part of a neighboring town. There will be 200 new service connections. The new customers will pay three times the regular city customer's water rate, to cover capital cost, maintenance and future expansion. If there are 800 regular city service connections, and these pay $1.50/1000 gallons for water, what percent of the city's annual revenue is contributed by the new customers? (Total water use is .4 MGD)

8. A water utility installed a compound meter at a local industry. On an average day, the turbine register reads 858,200 gallons. The positive displacement register reads 52,000 gallons. What percentage of the flow is on the small side? Was this meter a wise choice?

9. Monthly water rates for customers of the city of Rockville are as follows:
       $1.00/1000 gallons for first 200,000 gallons used;
       $.80/1000 gallons for the next 500,000 gallons used;
       $.50/1000 gallons for everything in excess of this.
    If the meter totalizer at Acme Industries registers 500,000 cubic feet of water
    used this past month, what is Acme's water bill for this period?

10. A utility with 125 miles of water mains which produces an average flow of
    12.12 MGD, has 8.71 MGD of metered customer sales, .37 MGD of metered
    use at public buildings, and 141 hours of firefighting/year at 2000 gpm. On the
    average, meters are underregistering by 2%.
    A. What is the percent lost water? (unaccounted for)
    B. Suppose unmetered public use is estimated at 1.6 MGD; recalculate lost
        water.

11. An unmetered water district has a nighttime flow of 1200 gpm and a daily flow
    of 2000 gpm. There is one industry within the district with a daily use of 800
    gpm and a nighttime use of 600 gpm. There is a 40 ft. diameter water storage
    tank within the district in which the water level is rising at the rate of 2 ft. per
    hour at the time of the nightly flow. What is the nighttime domestic use?

12. An industry moving into the city will require 3.8 MGD to process their product
    and adequate line pressure of approximately 60 psi at entrance. The water
    utility will provide the new service line (C=150) from the street main, which
    has a pressure of 60 psi; pipe and associated valves will be equivalent to 1000
    ft. of length. There is a 30 ft. drop in elevation from the main to the industry
    entrance. What is the minimum size service line which could be installed so that
    there will be no pressure drop from the main to the industry?

13. A new trailer court is to be metered. The trailers will all be rental units, and
    there is maximum space for 130 units when the court is full. Expect 10 rentals
    the first year; one is in and occupied now. Expect many trailers to be occupied
    by senior citizens who will fly south for the winter. Fire protection must be
    provided. One meter installation will service the whole establishment. Nearest
    water main size is 12 inch diameter, with a service line extending to the court of
    6 inch diameter. Required fire flow will be 1000 gpm. What meter installation
    will provide accurate flow readings for both high and low flow conditions for
    now, and for future years when the court is full?

14. An ice cube factory is moving into the city. It will need a 15 gpm flow for the
    ice cube machine, and 25 gpm for the water cooled compressors. There will
    also be a restroom and several hose connections. A 2 inch diameter service line
    services the building. A second ice cube machine may be installed in a few
    years. What meter should be supplied to meet present and future water needs?

15. The city golf club wishes to install an irrigation system, which will be metered. The service line size is 6 inch diameter, and it extends to a very small existing meter house. It is impossible to install a standard meter and backflow preventer and remain within the structure. The customer is very concerned about the pressure drops through the meter and RP device because sprinkler heads are at maximum spacing, requiring repumping to increase coverage area if pressures are reduced. The only length of straight pipe is a 9 ft. vertical pipe from the lower level of the building. The lower level is not accessible. (Max. flow = 850 gpm; Min. flow = 100 gpm). What type of metering assembly should be installed to pick up the small flows (leaks), provide low head loss at high flows, and still fit in the confined space of the existing meter house?

16. An apartment building will be constructed (single story) to house 20 families, on level terrain, at a distance of 300 ft. from an 8 inch diameter supply main with a pressure of 80 psi. Size the service line to the apartment, and install the meter.

17. An electroplating outfit in Florida which required 200 gpm for the process is purchasing a building which is 200 ft. from the nearest main. The industry will require at least 50 ft. of 1.5 inch diameter piping inhouse for processing. It may expand in a year or so. It also has a bathroom. Main pressure is 65 psi. Size the service line to the building, and install the meter.

CHAPTER 12

CENTRIFUGAL PUMPS I

Man has always needed to move water from one place to another against the forces of nature. Gravity will move it downhill on a grade. If depth is built up behind, that pressure will move it further. But when pressure is dissipated by friction loss, when water in the valley is needed on the mountain, the energy to move that water must be created. A pump is needed.

SCREW PUMP

Archimedes invented the screw pump in 287 BC. It physically lifts volumes of liquid, but has space and height limitations.

It is said that the Roman emperor, Nero invented the piston pump at about 100 AD.

PISTON PUMP

Volume after volume of water is displaced with each stroke, but this is a high energy consumer, and size limits capacity.

Pumping technology was restricted to these, and variations, until the nineteenth century. The basic concept of the centrifugal pump existed during the 1600's, but development was slow because the high speed drives needed were not available for another 200 years. It wasn't until the 1800's that the first fully functional centrifugal pumps were developed. These can move great volumes of water with much smaller units than the pumps previously in use. Centrifugal pumps operate with high efficiency, few parts are required, initial cost is low, and maintenance is relatively easy.

Operation of the centrifugal pump is based on the principle that a high velocity is imparted to the water as it enters the pump, and that velocity is converted to pressure as the water exits the pump. From a mechanical standpoint, water enters through the center, or "eye", of a rapidly rotating impeller. It is moved through the impeller vanes and thrown off their tips with great velocity. Still contained by the pump casing, the water cannot escape the unit, and collects at the "volute", or exit, slowing down as it leaves the pump.

From a hydraulic standpoint, note the energy changes that occur in the moving liquid. As water enters the pump from the suction piping, pressure is low. Passing through the impeller eye (smaller area), water velocity increases. This depresses the pressure even more, and acts to keep

water coming in from behind, because of the relatively higher entrance pressure. As the water comes into contact with the rapidly spinning impeller vanes, velocity is increased tremendously. This is the artificial energy which the mechanical action of the pump imparts to the water. It is Velocity Head energy which is increased at this point. The water is then rapidly thrown off the impeller tips against the pump casing, where it begins to slow down, collect, and then exit to the discharge piping through the only outlet, the progressively expanding volute. As Velocity Head decreases, Pressure Head increases; total energy must be maintained, and is merely converted from one form into another (Bernoulli's Theorem). The water now has enough new pressure to lift it the required distance.

The basic centrifugal pump working parts are impeller, casing, and the pump shaft, which connects it to the motor. There are no valves in the operating mechanism, and no close tolerances. Rotating parts do not touch stationary parts, and because of this, there is always some backward slippage of water between impeller and casing, which will vary depending on how much pressure the pump must provide.

Note: A centrifugal pump does not always pump the same amount of water.

In this respect it differs from a piston pump. However, a smooth constant flow is produced, which does not pulsate, as in the piston pump.

The design capacity of a centrifugal pump is determined by the diameter of the impeller eye, and the width of the impeller vanes. The amount of pressure which can be created is controlled by the diameter of the impeller, and the rotating speed (velocity of the impeller rim). The faster it moves, the more water velocity is created, the more pressure will be available as it slows down at the exit.

Pumps with closed impellers are most efficient, and are used for clean water pumping. Pumps with semi-open impellers are used for solids handling water. Pumps with open impellers are sometimes used to pump sludges. Obviously the more open impellers have a better chance of passing solids without clogging.

Figure 12.1 - Vertical centrifugal pump.

pump types: position

Vertical Pump

The impeller shaft is in a vertical position. Well pumps are designed this way; impellers are at bottom of the well, and the shaft extends upward to ground surface. Submersible pumps, with the motor close coupled to the pump, are vertical pumps used for underwater duty.

Horizontal Pump

Impeller shaft is in a horizontal position. The typical end suction pump has this design.

pump types: exit flow direction

Radial Pump

Flow leaves the impeller at a 90 degree angle from the direction it enters the pump. Most centrifugal pumps are radial pumps, they are able to produce high pressures. The High Service Pump at a potable water treatment plant, which lifts water from the plant to elevated storage is usually a radial pump.

Axial Pump

Water enters and exits the pump on the same plane. Backwash pumps - and Low Service Pumps, which carry water from the source to the treatment plant, are usually axial pumps.

pump types: suction condition

Single Suction

One suction line enters the pump casing.

Double Suction

The suction line splits in two as it enters the pump and water approaches impeller from both sides. This type minimizes head losses inside the pump. Double suction pumps are often seen as water distribution booster pumps.

Figure 12.2 - Double suction centrifugal pump.

pump types: impellers

Single Stage

One Impeller is installed on the pump shaft.

Multi-Stage

Used where water must be delivered at high pressures (a high pressure pump); to feed boilers, membrane processes, and for deep wells. It is designed so that the discharge from one impeller is led through a passage in the casing to the suction of the next impeller.

other important pump components

Stuffing Box and Packing

As the shaft extends to the outside of the pump, water is prevented from escaping by the stuffing box, which is filled with rings of packing. The packing acts primarily as a seal, and also as a lubricant for the shaft. It is pressurized by an external source of water at a higher pressure than the water inside the pump. Often this is a small flow of the pump discharge water which is directed back to fill the packing. This water pressure is increased because it has already passed through the pump. Pump sealwater flow is small, but constant, and stuffing box is designed to allow a constant leakage.

Mechanical Seals

This is an alternate to sealwater and packing, and has two components:

Rotating Assembly: attached to the pump shaft.

Stationary Assembly: positioned permanently on the stuffing box. The face of the stationary assembly provides a running surface for the face of the rotating assembly. There is no water leakage from a mechanical seal.

CENTRIFUGAL PUMP-CROSS SECTION

Bearings

Bearings take up the stress of the rotating shaft; they hold it steady, and support the impeller. The rotating part of the bearing is locked to the shaft; the housing holds the stationary side of the bearing. There are two types, both of which are in most pumps. However, there is great variation in bearing position, type and number, depending on the pump.

Radial Bearing (line bearing): this one is closest to the pump. It rides free in its own compartment and takes up and down stress.

Thrust Bearing: farthest from the impeller, closest to the motor. It takes up the major thrust of the shaft, which is opposite from the discharge direction.

Figure 12.3 - End suction centrifugal with stuffing box.

pumping characteristics:
the pump performance curve

For a given centrifugal pump, the total pressure developed, the horsepower required to drive it, and the resulting efficiency, vary with the discharge. The hydraulic principles which we are familiar with in systems where the pressure source is an elevated tank of water, also apply to pumping. Pressure and flow vary with each other. A pump is installed because pressure is low or nonexistent. Its purpose is to add pressure to the system - to lift the water and get the job done. This "head" which the pump must provide, or which it must "pump against", is dependent upon the requirements of the particular system. The pump must lift water up the hill, and it must also overcome head loss in the piping while it is operating.

> **Note:** A centrifugal pump will provide enough pressure to overcome lift and losses.
> No more!

Since every installation is different, yet not every pump is different, the flexibility is in
the flow. If the lift up the hill is a great distance, or if piping head losses are great (or
both), the pump will still do the work required; it will produce enough pressure, but will
operate at a much smaller flow than if the lift and/or losses were smaller. The same is
true when pumping a heavier liquid than water. If the Specific Gravity of the pumped
liquid is greater than one, the pump will lift it the required distance, but the flow will be
less than if it were pumping water.

Recall the lack of close tolerances between the impeller and pump casing, which allows
backwards slippage of water as the pump operates. This slippage is greater as the head
(lift plus losses) increases, and therefore, flow decreases.

Horsepower requirements also vary with flow. Generally, if the flow is greater, the
horsepower required to move the water will be greater.
Pump performance is measured in gallons per minute delivered, height to which water
is lifted, and efficiency. Each pump manufactured has a designated amount of head and
flow at which it will operate most efficiently, and pump purchase should be based on how
close to peak efficiency it will come in a particular system. This allows the pump to
operate with the least strain, both mechanically and hydraulically. A pump which
operates well off peak efficiency may do the job, but results will be excessive energy
requirements and shortened pump life.

The interrelations of pump head, flow, efficiency and horsepower are known as the
Characteristics of the Pump. These are important elements in pump performance, and
they are diagrammed graphically on a Performance Curve. The curve puts these
elements together to show a picture of what the pump can do. This is the Pump
Characteristic Curve, and each centrifugal pump will be expected to perform as its curve
describes. It will not stray from this.

If flow from a pump is measured, head (lift and losses) is calculated, and the
characteristics do not match the curve, it is most likely that the losses have been
estimated incorrectly.

PUMP CHARACTERISTIC CURVE

Inspecting this typical Pump Characteristic Curve:

The horizontal scale on the bottom represents the flow (gpm) which this pump will deliver. This is plotted against head, and creates the H-Q curve. This one is the base for most pump curves. The vertical scale represents head, in feet of water. Head is composed of lift plus losses which will be overcome by this pump. It is the total amount of useful energy that will be transferred from the impeller blades to the liquid. This is <u>Total Dynamic Head</u>, and it registers visibly as the difference between the suction and discharge gage pressures.

From the horsepower scale on the right is plotted power consumed, also against flow, and comprises the P-Q curve. This is <u>Brake Horsepower</u>, and refers back to Chapter 5, where power calculations were introduced:

$$\frac{gpm \times hd}{3960} = \frac{whp}{pump\ eff.} = bhp$$

The pump curve is actually derived from this formula. Try it. Pick any head on the curve. Read flow, and efficiency at that flow. Calculate brake horsepower with formula. Then look it up on the curve. It will match.

The Efficiency curve, E-Q, is also plotted against flow. Because of internal losses within the pump itself, no pump is 100% efficient. New pumps are usually about 85% efficient, and efficiency decreases with age.

Note: All the curves are plotted against flow; when reading the curve determine flow first, then proceed up from flow at the bottom to the curve and element desired.

The Rated Head and Flow are those at which maximum efficiency is reached. This is the Design Point, the point at which this pump was meant to operate. The pump is purchased because the head and flow conditions at this pump's design point matches the head and flow demands of the system.

Changing System Conditions

Each pump has a maximum pressure that it can achieve. If the pressure needed is greater than that which the pump can deliver, the pump is said to be running at Shutoff Head.

With a pump having this characteristic curve, assuming no friction losses, the pump would lift the water 120 ft. up the pipe. The pump would keep running, but the water level would remain there. The pressure of the discharge water in the pipe would be equivalent to the maximum pressure that this pump can achieve, and so there will be no flow. The water inside the pump will heat up which will result in damage to the pump. This would be equivalent to operating the pump with the discharge valve closed. Horsepower requirements are lowest at this time; the water isn't going anywhere, just circulating around inside the pump casing.

It is, however, often desired to start up a centrifugal pump in this condition, with the discharge valve closed when horsepower requirements are lowest, in order to start the flow moving slowly. The valve is then gradually opened; this minimizes surge upon startup, and gives the water in the suction line a chance to overcome inertia and get moving into the pump.

Inspecting the characteristic curve again, extend the H-Q curve down farther than it is drawn, on the right. The pump would be running at high flow and negative head. This can happen upon starting a pump; a huge initial flow cannot overcome the inertia of the water in the suction pipe, and the water will split, or pull away from itself, creating a vacuum and drawing air into the pump. Centrifugal pumps have great difficulty expelling air, which may cause damage to the pump. It is better to start up with the discharge valve closed, or have the valve automatically operated so that it opens slowly as the pump starts. This phenomenon of high flow/negative pressure should be familiar. The same thing occurs in a water line when flow suddenly increases, upon a main break, or a hydrant fully and suddenly opened: large flow at once, vacuum created upstream, water separates. This invites a chance for backflow of cross connected contaminants into the system.

Take another example:

The pump is installed to discharge water horizontally through a pipe. There is no lift required, so the pumping head will be equivalent to the head losses only. The flow delivered will be that which matches up on the characteristic curve with the system's head losses.

When a partially closed valve is added to the end of this pipe, the flow delivered will be much less, because minor head loss created by the valve is added to the friction head loss in the pipe. The Total Dynamic Head to be achieved by the pump is greater; therefore flow will decrease. Refer back to Chapter 6 for Friction Losses in Fittings chart.

When the valve is totally closed, no water flows. The pump overheats. The same situation would exist if the valve were removed and the piping were extended to the point where the head losses in this horizontal line were greater than the pump could overcome at any flow. The pump would be operating at shutoff head, and the water would only go so far down the pipe, and then stop. For this situation a pump with a higher shutoff head (a stronger pump) is needed, or a booster pump could be installed midway down the pipe to raise the pressure again.

On the other hand, if the head on the system should drop significantly for any reason (break in discharge line, perhaps), the pump would deliver a greater volume of liquid, and might easily reach the point of maximum horsepower requirement, and damage the motor.

If the needs of the system were expected to change in this manner, and a high flow were desired at intervals, from the same pump, then purchase of a larger pump would be in order, to fulfill this system requirement. To maintain best possible efficiency under all conditions, a pump with a flat H-Q curve would provide less variation in head over a greater capacity range. Sometimes changing to a wider impeller, or one with more vanes, will create the same effect.

If the condition of head-too-low, flow-too-high should exist at all times in a pumping installation, and be undesirable because this causes the pump to operate off peak efficiency, then running the pump with the discharge valve slightly closed will put more head loss into the system and decrease the flow, bringing operation nearer top efficiency.

The System Curve

In choosing a pump, it is desired to match the design point (head, flow) of a particular pump's curve with the conditions of the system. The system is the arrangement of pipe, fittings, and equipment through which the water will flow. The easiest way to match H-Q characteristics of the system with H-Q characteristics of various pump curves is to develop a System Curve. The system head consists of lift plus losses. Lift is set, and can be measured, but head losses depend upon the amount of flow being pumped. How can the effect of system head loss on the pump's flow be determined, when head loss cannot be calculated until after the flow is known?

If a system curve is developed, with calculated heads based on a range of flows, this will represent the characteristics of the system, the actual conditions under which a pump will operate. Then the H-Q point at which the pump should operate can be chosen, and marked. Inspecting several possible pump curves (representing several different pumps), they can each be superimposed onto the system curve, and the one that matches the desired point, at a place on the pump H-Q curve which is at or near its design point, that is the pump which should be purchased.

To develop the system curve, choose a flow lower than what is should operate at, and calculate the head losses at that flow (Hazen-Williams); add this to the required lift, and mark it as "head" on the graph. Do this with several different increasing flows, and draw the graph from point to point. This curve now represents the characteristics of the system. Note that system curves do not usually start at the bottom of the graph. The curve itself represents only the head losses. The level it starts at represents the lift.

If it is desired to operate the system with more pressure in the lines than that which a pump will provide with the system as is, then theoretically add some more static head only to existing lift, and develop the graph with this included. After the pump is purchased and installed, a valve will have to be installed at the far end of the system to add this extra loss, and bring the flow into the range planned with the curve.

The system head includes both suction and discharge piping. Keep in mind that static head may vary, if dependent on tank drawdowns and filling. Both maximum and

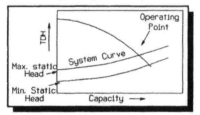

minimum static heads should be used to develop the system curve. The distance between them will be the actual operating range of the pump chosen. As a short cut, head loss through pipe and fittings may be calculated for one flow, and losses for other flow rates determined from this relationship:

$$\left(\frac{Q1}{Q2} = \frac{H_1}{H_2}\right)^{.54}$$

Pumps in Series

When system heads are too great for one pump to overcome, pumps may be placed in series, one after another in the same pipeline. This puts more than one source of pressure into the system, and it <u>increases the head</u>. However, these pumps are pumping the same water through the same system. <u>Series pumping is not meant to increase the flow</u>.

Booster Pumping: the system has expanded; pipelines have been extended, or an apartment building has been added; more pressure is desired to lift the water that extra distance. A booster pump is installed at a low pressure area to raise the head to a safe and practical level for customer use. When choosing a booster pump for the system, take the pump curves

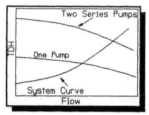

from the original and the booster. At the desired flow, add the heads of both to arrive at the operating head with the series arrangement. If the original source of pressure is an elevated tower, this is still considered a series pump arrangement; use the depth of water in the tower as the head for that unit. For a diagram of this effect, refer to the graph including the system curve transacting pump curve - with one pump installed, and then with two pumps installed.

Multi-Stage Pumps: when impellers are in series in the same pump casing, it is hydraulically a series pumping setup. High pressure pumps and deep well pumps must achieve great heads. The discharge of one impeller is directed into the eye of the next impeller. The flow is constant throughout, but at each stage the head increases; each impeller is considered a separate pumping unit with a separate curve. Heads are additive, and total head is limited only by the efficiency of each. Identical impellers are used.

In most series pumping arrangements, identical or similar pumps are employed. To project horsepower use with pumps in series, use the total head to be achieved, and calculate. Efficiency can be calculated by multiplication of the efficiencies of each.

Pumps in Parallel
When head is sufficient, but more flow is needed, pumps may be arranged in parallel. Wastewater lift station pumps, and treatment process pumps are placed in parallel to cope with the constantly fluctuating flow. Industrial processing needs specific volumes of water for washing, rinsing, processing a product; parallel pumps will be in use.

From the standpoint of each pump in this sketch, the system requirements are the same, and must be met. Each pump achieves the same head for its flow, but since there are two of them, the system is getting twice as much water.

Pumps arranged in parallel increase the flow, but the head remains that of one pump working. When combining pump curves to determine characteristics, the combined discharge is the sum of the discharges of both pumps at that head. Once again, the curve with the system imposed should be referred to:

Parallel or series operations allow the operator to be flexible enough in pumping capacities and heads to meet requirements of system changes and extensions. Multiple pump operation also allows one pump to be taken out of operation for repair without shutting down the entire system.

Figure 12.4 - Wastewater influent pumping - a "pumps in parallel" situation.

PROBLEMS

Many of the problems for this chapter refer to one of the five typical pump characteristic curves which are included following the list of problems.

1. Referring to pump curve #1:
   A. When the pump is operating under 90 ft. of head, what is the horsepower use?
   B. What is the head at which this pump will run most efficiently?
   C. What is the shutoff head for this pump?
   D. When this pump is operating at 30% efficiency, what two heads is it pumping against?

2. Referring to pump curve #2:
   A. At what two heads will this pump operate at 70% efficiency?
   B. This pump is designed to work at 130 ft. of head. Working at this design point, what flow is achieved?
   C. If the metered flow from this pump suddenly increases from 1200 gpm to 2200 gpm, what is the new head it is pumping against? What would you suspect caused this?

3. Given a pump with the characteristics of pump curve #1, what is the delivered flow under each of these conditions?
   A. Discharge lift 100 ft.; headloss in discharge piping 15 ft.; suction head 25 ft.; suction head losses 10 ft.
   B. Discharge lift 100 ft.; discharge piping head loss 35 ft.; suction lift 15 ft.; suction head loss 5 ft.
   C. Discharge lift 50 ft.; headloss in discharge piping 12 ft.; suction lift 20 ft.; suction head loss 13 ft.

4. Referring to pump curve #1:
   The pump has a static suction head of 15 ft., and is pumping 200 gpm from a tank. Total friction head loss in the system is 20 ft. What is the elevation difference between the pump and the end of the discharge line?

5. Referring to pump curve #1:
   A. What is the maximum percent efficiency this pump can achieve?
   B. What is the motor horsepower needed if the motor is 80% efficient and the pump is delivering 140 gpm.
   C. If the brake horsepower is 8 hp, what head is the pump achieving?

6. A pump delivers 100 gpm to an elevated storage tank 50 ft. above the pump. Head losses in suction and discharge lines are 12 ft. total. Suction lift is 15 ft. The pump is 76% efficient and the motor efficiency is 85%. Power cost is $.15/Kw hr. What is the daily pumping cost?

7. A pump draws 60 Kw. Head loss is 9 ft.; velocity is 5 ft./sec. Pipe is 3 inch diameter. If the wire to water efficiency is 60%, how high can the pump lift water in this system?

8. Below are the characteristics of a centrifugal pump:

Flow (gpm)	Total Head (ft.)	Efficiency (%)
200	124	30
400	123	54
600	119	67
800	114	76
1000	107	83
1200	98	84
1400	85	83
1600	73	76

Plot the curves showing the operating characteristics of this pump. What is the design point of this pump?

9. Construct a horsepower requirements curve for pump curve #4.

10. Referring to curve #5:
The electricity bill this past month was $5,000.00. The flowmeter on the pump records 800 gpm, and electrical cost is 10 cent/Kw hr. How efficient is the motor?

11. Referring to pump curve #5:
The pump lifts water 240 ft. and develops 27 ft. of friction head loss in the pipe. If the motor is 90% efficient, and electrical costs are $.08/Kw hr., find the cost to run this pump for a month under these conditions.

12. Referring to pump curve #1:
A. At what head will this pump draw minimal horsepower?
B. If the pump were operating with a suction lift of 12 ft., and discharging through 4000 ft. of horizontal line with a slope of .015, what would be the flow?

13. Water is pumped from a storage tank through a 4 inch diameter steel pipe. The velocity in the suction line is 6 ft./sec., and the distance between the pump and the storage tank is 20 ft. The pump discharges through a 3 inch diameter cast iron pipe 100 ft. long to an overhead tank. The water level in the overhead tank is 70 ft. above the level of the solution in the storage tank.
A. What is the pressure that the pump must develop? (psi)
B. If this pump is operating at 60% efficiency, what is the brake horsepower?

14. If a tank 150 ft. long, 10 ft. deep and 35 ft. wide is being dewatered with a pump rated at 100 gpm, and it takes exactly three days to empty the tank:
    A. What is the actual pumping rate?
    B. How efficient is the pump?

15. A wastewater pumping station has a static discharge head of 73 ft. The force main is 8 inch diameter DIP, 4600 ft. long. Minor station losses are estimated at 5 ft., and the flow rate is 600 gpm. What Total Dynamic Head are the pumps achieving? (C=120).

16. A pump delivers 1500 gpm through a 12 inch diameter pipe. How many 6 inch diameter pipes of the same length would be required to carry the same volume from the pump?

17. A low lift pump brings water from the lake to the treatment plant, which lies at the same elevation. System characteristics are:

    Suction Line:
        60 ft. of 4 inch diameter new DIP
        one open gate valve
        four 90 degree long sweep elbows
        reducer 4 inches to 2 inches

    Discharge Line:
        one open gate valve
        one open globe valve
        three 90 degree elbows - long sweep
        reducer 4 inches to 2 inches
        4 tees using through run
        1800 ft. of 4 inch diameter new DIP

    What is the water horsepower required to deliver 200 gpm to the treatment plant?

18. Water is pumped up 15 ft. from a tank through 30 ft. of 10 inch diameter new steel suction piping with a gate valve and a standard elbow - to a pump which has the characteristics of curve #5, and then discharged through 5000 ft. of the same diameter pipe past a check valve, a gate valve, two standard elbows into another tank 250 ft. above the pump. What is the flow from the pump?

19. Water flows through a main at a rate of 200 gpm. A pump with the characteristics of curve #1 is installed to take water from a tank and add to the flow in this main. The pump has a static suction lift of 10 ft., and a 4 inch diameter horizontal discharge line 200 ft. long. The main is 8 inches diameter and 3000 ft. long. What will be the flow from the pump?

20. A ductile iron pipeline 1000 ft. long will connect two reservoirs. One is 150 ft. higher than the other. A pump will be installed to pump water from the lower one up to the higher one at a rate of 500 gpm. If a pump with the characteristics of curve #2 is chosen, what should be the diameter of the pipe which connects the two reservoirs?

21. Given two identical pumps, each with characteristics shown in curve #2, what is the discharge for a head of 140 ft:
    A. If both pumps are operated in series
    B. If both pumps are operated in parallel

22. When a 3 stage well pump with characteristics as in curve #3 is in use - pumping from a depth of 1000 ft. through a 12 inch column - to elevated storage which is 80 ft. above ground - over an above ground discharge distance of 200 ft. What is the discharge of the pump?

23. The city is supplied with 1 MGD of potable water. Over the past few years, the city has increased population by 50%, and has expanded to the suburbs. Water storage was previously provided by a ground level tank; pressure from this tank is now inadequate for the outlying areas. An elevated tank 2 miles from the treatment plant will now be installed to replace it, and 80 psi will be available at the new tank.
    A. How high should the elevated tank be built?
    B. What horsepower high service pump will be needed?

24. An industry requires 400 gpm from a river 950 ft. away. The industry is 41 ft. higher in elevation than the river.
    A. Design the pumping system and size the pump to be purchased. Include fittings.
    B. If you had to choose from among the 5 pumps whose curves are given, which one would you choose?

25. A consulting engineering firm has been contracted to design a sewage lift station to move water into the treatment plant. The station must handle a 2 MGD flow. Pipe length is 700 ft. Vertical lift is 60 ft. Submersible pumps are desired. Must design for now and for the future. You are the consulting firm. Design it.

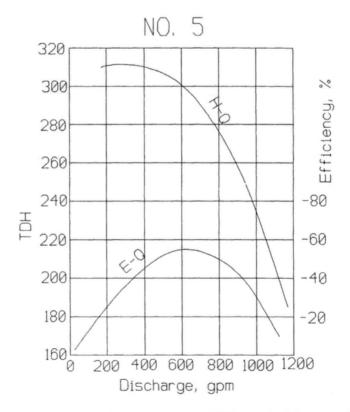

CHAPTER 13

CENTRIFUGAL PUMPS II

Once a pump has been installed, chances are that it will remain in operation whether or not performance is as expected. Pumps are sometimes purchased oversized, with expected duty to be heavier as system needs increase. A pump which is too powerful for the present system does not last longer. The pump curve will not match the system curve and operation will be off maximum efficiency. On the other hand, a pump operating at design rating when first installed, will not be as efficient as system needs change with time.

changing pump performance: pump affinity laws

The pump itself has some flexibility for change. Pump characteristics can often be adjusted to meet the system, and yet maintain peak efficiency. Each pump characteristic curve applies to that pump with the provided impeller, operating at the stated speed; impeller size and pump speed are printed on the nameplate, but in many pumps they can be changed. Changing the rim speed of the impeller may be the most economical way to make adjustments; it will predictably affect head, flow and brake horsepower. This actually changes the rating of the pump and develops an entire new pump characteristic curve for it.

Pump speed may be changed in either of two ways:

Change pumping RPM: a fixed gear drive - adjusted up or down will change shaft speed. Extra space is required for this equipment.

Change impeller size: many pumps can be fitted with a number of different sized impellers. With a smaller impeller the pump shaft will turn at the same rate, but the velocity at the rim of the impeller decreases. Not all pumps are originally purchased with the largest sized impeller installed. Many start with a smaller size, which better matches system characteristics at the time, and change to a larger impeller later, as the system ages.

The laws governing the effect of varying pump speed are known as <u>Pump Affinity Laws</u>, and are as follows:

When speed is changed, capacity varies directly as the speed, head varies directly as the square of the speed, and brake horsepower varies directly as the cube of the speed.

To simplify and explain:

$$\frac{\text{New RPM}}{\text{Old RPM}} \times Q = \text{New Q}$$

$$\left(\frac{\text{New RPM}}{\text{Old RPM}}\right)^2 \times \text{Head} = \text{New Head}$$

$$\left(\frac{\text{New RPM}}{\text{Old RPM}}\right)^3 \times \text{bhp} = \text{New bhp}$$

If impeller diameter is being changed, the effect is the same.

$$\frac{\text{New Diameter}}{\text{Old Diameter}} \times Q = \text{New Q}$$

$$\left(\frac{\text{New Diameter}}{\text{Old Diameter}}\right)^2 \times \text{Head} = \text{New Head}$$

$$\left(\frac{\text{New Diameter}}{\text{Old Diameter}}\right)^3 \times \text{bhp} = \text{New bhp}$$

Pump curves are available which demonstrate characteristics using different size impellers. They are called Oak Tree Curves and are a way of demonstrating the performance range of the pump. The efficiencies appear on the curve like rings of an oak tree. The affinity laws show theoretical expected performance, but the curve shows exactly how the pump will operate.

Sometimes the width of impellers can be changed. A wider impeller throws a greater volume of water, and usually produces a flat Head-Capacity curve. Narrower impellers pump less and give a steeper curve. Impellers with more vanes produce a flatter curve also. With less vanes a steeper curve is produced. The size of the impeller eye can also increase or decrease pump capacity.

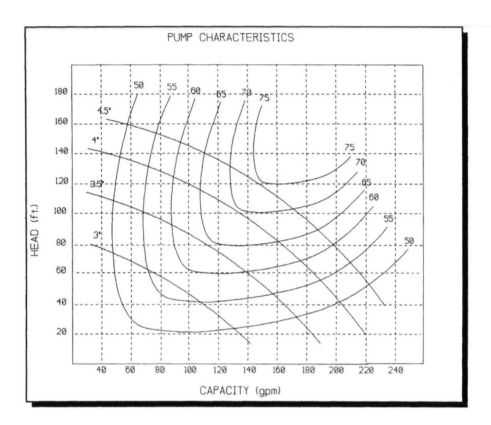

PUMP CHARACTERISTICS

variable speed pumps

For a system whose needs change with operation a variable speed pump may be the best choice. The speed of the pump will change to match the system needs. For instance, if the system must alternately deliver flow at different capacities, or if static lift is small, and head losses are great and constantly changing, the pump head of a variable speed pump will vary with system head and the pump chosen will operate efficiently at several different capacities.

Advantages of using variable speed pumps are in the energy savings obtained by operating close to peak efficiency regardless of changing system needs, and in the ability of the pump to discharge less water when demand is for less, with resultant smaller horsepower demand. Variable speed pumps are commonly used for wastewater pumping, especially at lift stations, where flows vary widely. They are good choices when the average quantity of liquid being pumped is less than two thirds of the maximum design flow and operated in response to information received from tank level controls; pump speed will be what is needed to keep the proper pressure differential.

Disadvantages of variable speed pumps are high cost, greater maintenance, and reduced overall efficiency.

priming a pump

A major deficiency of centrifugal pumps is that the pump chamber must be filled completely with water upon startup for it to function correctly. If there exists a suction head (positive pressure on the suction side of the pump), the unit will always remain full whether on or off, but with a suction lift, water tends to run back out of the pump and down the suction line when the pump stops. If the casing is filled with air or vapor, the impeller cannot create enough vacuum upon starting to draw water back into the unit, and the gas will just circulate around in the pump. The heat produced by the pump's mechanical action has no flowing water to dissipate it, and both pump and motor will overheat in a short time. It is best to always provide a positive suction head for a centrifugal pump, but if that is not possible, the pump must be separately primed (filled with water) each time it is started. Rather than manually filling the pump with every startup, several different types of self-priming devices have been developed.

Foot Valve: a specialized check valve at the end of the suction line. It rides open when the pump is operating but shuts as the pump stops, trapping water in the unit and piping. Recorded efficiency of foot valves varies; if the valve doesn't seat properly, water will leak out.

Vacuum Pumps and Ejectors: separate units attached to the main pump casing, they create the vacuum needed to fill the pump before startup. These are commonly seen installed with low lift pumps at water treatment plants; the disadvantage is extra capital cost and maintenance.

Priming Chamber: a permanently fitted reservoir of liquid above the pump. It is usually filled by a small recirculated flow of the pump discharge water, and empties its contents as the pump starts. This is a bulky unit, which must be able to completely fill the pump casing and associated piping.

Some self-priming pumps are built with a wide suction inlet which rides higher than the pump. Upon stopping, the discharge valve closes, and the water in this suction area slides back into the pump, rather than down the suction line. A vent exhausts any extra air.

cavitation

Water naturally contains a small gaseous phase, present as tiny bubbles or vapor molecules, which are in constant motion and always trying to escape the fluid. The percentage of this gaseous phase which remains in the water without evaporating, depends on the water's tendency to evaporate. At normal temperatures in an open vessel, atmospheric pressure upon the water minimizes this tendency; the vapor areas are fairly stable, and the water remains in the liquid phase. At elevated temperatures, or under conditions of lowered air pressure, the water's tendency toward vaporization increases. At higher elevations water boils at lower temperatures.

As water enters a pipe and pushes the air out, this evaporative tendency of the liquid, becomes, by comparison, more significant. If the water is under water pressure, instead of air pressure, the vaporizing tendency is minimal, but only because normal inpipe water pressures are usually a good deal more than atmospheric pressure (at 50 psi, water will not boil until a temperature of 300 degrees F is reached).

This tendency to vaporize, or, the pressure that the water exerts on its surroundings, is called <u>Vapor Pressure</u>. At 60 degrees F, water has a vapor pressure of .256 psia - very small. At 212 degrees F, water has a vapor pressure of 14.7 psia, the same as the atmospheric pressure upon it, and it boils.

When the pressure of the atmosphere falls below the vapor pressure of the water,the amount of gaseous phase rapidly increases, bubbles are formed, and the water begins to flash into vapor. In a pipe, if the vapor cavity becomes so large as to fill the cross section of the pipe, a true column separation occurs. A water main break can cause this to happen. The increased flow creates vacuum, and some of the water flashes to vapor to fill the vacuum. If a valve is closed too rapidly, upstream of the valve pressure suddenly increases, causing hammer, but downstream of the valve, the Velocity Head keeps the water moving, vacuum forms in the space behind, and the water starts to vaporize in that area. Rapidly opening a valve can have the same effect on the upstream end. Power failure to a pump also will cause negative pressure waves to be set up which will travel downstream and produce vapor formation.

These pressure changes are momentary; as pressure equalizes in the system, and returns to the area, the low pressure condition is eliminated. The column rejoins, and the vapor bubbles collapse, or implode, quite forcibly, against the surface of the surrounding metal, chipping and disintegrating that metal, bit by bit. This phenomenon is referred to as <u>Cavitation</u>.

The effects of cavitation are most damaging in areas where it is occurring repeatedly. In pumping operations employing a suction lift, there is no positive depth of water to provide an initial pressure to the water entering the pump, and it is here that cavitation is most significant. As a pump impeller rapidly increases water velocity, pushing fluid out of the pump, a tendency toward vacuum is created in the water coming in; in effect, a space is left between molecules of incoming water, and pressure is low. There is only atmospheric pressure on the intake reservoir which pushes the water in to fill the suction pipe and take up these spaces. Since atmospheric pressure at sea level and standard temperature is equivalent to 34 ft. of water pressure, a pump feeding from a tank whose surface elevation is at the same level as the pump centerline, really has a positive suction head of 34 ft. of water, as absolute pressure. Theoretically the suction pipe could be lowered to 34 ft. below the pump, and the liquid will still rise into the pump. This amount of absolute pressure needed to get the water into the pump is called Net Positive Suction Head (NPSH). It is the total amount of energy, in Absolute feet of water, at the pump centerline.

If this Net Positive Suction Head (theoretically 34 ft. or 14.7 psia) should drop below the vapor pressure of the water entering the pump, the external pressure will become less than the water's tendency to vaporize, and the water will start to flash into vapor at the pump entrance, eye, and impeller vanes. Bubbles will form, which will be forcibly collapsed as the water pressure suddenly increases near the pump exit. Cavitation will occur, with resulting chipping and destruction of internal pump parts. Operated this way for a length of time, the vibration caused will also damage bearings, shaft, and seals. The destructive effect is magnified further because the water molecules turning to vapor expand many times in the phase change, which enhances the repressurization and implosion.

But how does this theoretical 34 ft. of water pressure (14.7 psia) become less than the theoretical vapor pressure of water (.59 ft., .256 psia)? It does happen, because these are both only theoretical, and static quantities. There are friction losses within the mechanism of the pump itself. As the water enters the impeller eye, it makes a right angle turn. Turbulence is increased by the vanes, adding to head loss. The eye, and the vanes have a degree of roughness, etc., all sources of head loss. Recognizing these internal pump losses, manufacturer's data will supply a specific amount of minimum positive absolute pressure which must be present in the water at pump entrance in order to overcome these head losses, keep the pump running efficiently and without cavitation. This is the Net Positive Suction Head Required (NPSHR). It is a characteristic of each pump, and is obtained by calculations and testing of the prototype pump of each model at a range of flows. It is determined by shape, size, and speed of pump; it increases with increased flow, and is often inserted as a NPSHR curve on the pump performance curve.

NPSHR is the amount of absolute pressure needed to use the pump. Now it must be determined how much pressure the system will provide. This amount is the Net Positive Suction Head Available (NPSHA), a characteristic of the system. It is the energy actually available, in feet of Absolute Pressure, at the inlet of the pump. NPSHA is calculated for each installation with a simple formula:

NPSHA = (Atmospheric Pressure + Velocity Head) - (Lift + Losses + Vapor Pressure)

           positive pressure forces   negative pressure forces

---

Note: All of the elements of the formula are recorded in Absolute feet of water.

Atmospheric Pressure: atmospheric pressure may be 14.7 psia at sea level, but it is substantially lower in Denver, Colorado, a mile above sea level (see chart); there is less positive pressure to start with.

WATER CHARACTERISTICS TABLE			
Altitude Below or Above Sea Level	Atmospheric Pressure Pressure PSIA	Equivalent Head of 75° Water Feet (Abs)	Water Boiling Point °F
- 1000	15.2	35.2	231.8
- 500	15.0	34.7	212.9
0	14.7	34.0	212.9
+ 500	14.4	33.4	
1000	14.2	32.8	210.2
1500	13.9	32.2	209.3
2000	13.7	31.6	208.4
2500	13.4	31.0	
3000	13.2	30.5	
3500	12.9	29.9	
4000	12.7	29.4	204.7
5000	12.2	28.3	
6000	11.8	27.3	201.0
7000	11.3	26.2	
8000	10.9	25.2	197.4
9000	10.5	24.3	
10,000	10.1	23.4	193.7
15,000	8.3	19.2	184

Velocity Head: when operating with a large suction lift, Velocity Head may be significant enough to count. Use the water velocity at the impeller eye. This is the area of lowest pressure, and usually where cavitation starts.

Suction Lift: it varies with the installation, and also may vary with drawdown if water is pumped from a tank. Use the lowest level.

Suction Losses: Head Loss will depend upon the flow, and will vary with it as the pump operates. Use the maximum flow the pump will be discharging at any time.

Note: NPSHA must be greater than NPSHR in order to prevent cavitation in the pump!

Vapor Pressure: varies with water temperature. For cold water flow efficiently pumped .59 ft. is the vapor pressure. However, with higher heads and lower flows the water heats up. Operating a pump with the discharge valve throttled almost to the point of shutoff head may heat up the unit enough to cause cavitation. Reducing flow by recirculating a percentage of discharge water back to the suction reservoir instead of throttling a valve may help, but even with this the same water is being pumped over and over again, heating up more with each round trip. Hot climates, overheated pumphouses, direct sunlight all contribute to the problem, increasing vapor pressure.

PROPERTIES OF WATER

Temperature °F	Absolute Vapor Pressure		Specific Weight lb./cu.ft.	Specific Gravity
	psia	ft. of water		
32	0.088	0.20	62.42	1.0016
40	0.122	0.28	62.43	1.0018
50	0.178	0.41	62.41	1.0015
60	0.256	0.59	62.37	1.0008
70	0.363	0.89	62.30	0.9998
80	0.507	1.2	62.22	0.9984
90	0.698	1.6	62.12	0.9968
100	0.949	2.2	62.00	0.9949
110	1.275	3.0	61.86	0.9927
120	1.693	3.9	61.71	0.9903
130	2.223	5.0	61.56	0.9878
140	2.889	6.8	61.38	0.9850
150	3.718	8.8	61.20	0.9821
160	4.741	11.2	61.01	0.9790
170	5.993	14.2	60.79	0.9755
180	7.511	17.8	60.57	0.9720
190	9.340	22.3	60.35	0.9684
200	11.526	27.6	60.13	0.9649
210	14.123	33.9	59.88	0.9609

A cavitating pump usually makes noise like rolling marbles, gravel grinding, or balloons popping. When excessive, it can be very loud and rasping. Sometimes it can occur with no noise at all, and perhaps is then most serious, for it goes undetected.

In summary, some of the factors which may drop NPSHA to below NPSHR, and start a pump cavitating are:

Excessive Lift
Excessive Head Losses
High NPSHR
Low Atmospheric Pressure
Hot Water

other causes of cavitation

Entrained Air: outside air may enter the system on the suction side. It isn't as damaging as vapor caused cavitation, because there isn't the expansion of vapor within the pump, but it can be more difficult to discover. Air can enter the pump through pump packing, suction valve packing, and from flanged joints and shaft sleeves. The velocity of the water moving through the pump's system does create a slight eductor effect, and can easily suck air into the system through an opening.

Insufficient Submersion: if the suction line is not submerged adequately in the suction tank, vortexing may be set up around the pipe, drawing air right in with the water.

Shallow Suction: a suction tank filling rapidly as the pump removes water, especially if it is waterfalling in, will trap excessive air, and may short circuit it right into the pump.

Slope of Associated Piping: if the discharge line is lower than the pump centerline, and the discharge valve is opened before the pump is started, vapor can escape backwards into the pump. The same can happen if the suction pipe is not designed with an inclined slope. Air pockets will form in the line, and as they build up, now and then flush into the pump. There should be 5-10 pipe diameters, or 3 impeller diameters from pump inlet port to nearest suction elbow. Increased turbulence at the elbow releases dissolved oxygen from the water, which may catch on the impeller. Highly aerated water will magnify this tendency.

Reducers: at pump entrance the suction pipe decreases size through a cone shaped fitting called a reducer, and at discharge increases size again by the same method. Concentric reducers are only used on the discharge side. Eccentric reducers should always be used at suction entrances. When the supply line enters from below, an eccentric reducer with flat side up should be used at suction, leaving no chance for air to collect in the pipe at pump entrance. When the supply line enters from above, (positive suction head), an eccentric reducer with curved side up should be used at suction. This gives any gas collected there a chance to slide back up the suction line away from the pump. In this pumping arrangement (positive suction head), there is no risk of cavitation from low

CONCENTRIC REDUCER

ECCENTRIC REDUCER

ECCENTRIC REDUCER

pressure conditions, but air pockets, no matter what the pressure condition, if becoming large enough inside a pump, can cause vapor lock, or air blockage. The air cannot be expelled and the pump's capacity will be greatly reduced (a pump filled 5% with air will reduce capacity by 50%). If the air pocket is large enough, no water will be discharged at all. A pocket of air in a pump, a pipeline, a sand filter, or any other water-filled container, is usually not expelled by the moving water. Rather, the water routes around it, reducing the effective working volume of the unit.

In the above sections reference was made to air becoming entrained in the system from the outside and to water vapor formations from within the liquid, but the same principles, and damage, can be applied to any gas which becomes trapped in pump or associated piping. In wastewater pumping it is likely that gases other than water vapor or air will be formed within the wastewater, and lead to cavitation. Wastewater pumps should always have a flooded suction, if possible.

Figure 13.1 - Pump reducer on discharge side.

cavitation in piping

Cavitation can cause stress in piping when velocity is high and direction of flow is sharply changed. Water going around an elbow or tee at high velocity will pull to the outside of the bend, creating thrust there, but on the inside of the bend, pressure may drop to a point corresponding to the vapor pressure of the fluid, causing bubbles to form, and then collapse slightly downstream as the flow straightens and pressure returns.

A similar phenomenon occurs at an ordinary pipe entrance from tanks. As the water rounds the bend and increases velocity exiting the tank, the flowlines cause a slight contraction of the stream at pipe entrance. This creates a low pressure area along the inside wall of the pipe where vapor forms and cavitation can occur.

specialized pumping systems

Submersible Pumps

Submersible pumps see increased service each year for many applications. Now commonplace for sewage and stormwater pumping, submersibles remove urban drainage water to and from retention basins, irrigate farmland and golf courses, and replenish wetlands. They

provide good low pressure pumping for transfer applications: pond to pond, from ditches, and for construction dewatering. In potable water service, submersibles provide high pressure for deep well pumping when equipped with multi-stage impellers.

Consider the wastewater lift station.

The double well system employs an end suction pump, which takes water with a positive head from the separate wet well, and pumps it through a force main to the surface. The dry well section of the installation houses pump, motor, control panel, and all associated equipment. This structure often extends well above ground as the pumphouse.

The single well system has a wet well only, with submersible pump installed at the bottom. The pump is submerged, but the control panel, including starters and alarm system for high water level, is above ground. The alarm may be visual, audible, or remote readout through telemetry to the treatment plant. The pump is installed on guide rails, can be easily hoisted out for service, and is equipped with a quick disconnect coupling for removal: a sealing flange automatically releases and re-engages the discharge piping. The submersible pump lift station can be built in place or prefabricated, and is sized by the inflow during the peak period of the day. The high water level is determined by the elevation of the basement most likely to flood, and the low water level is set by the elevation of the incoming sewer and the top of the pump casing.

Submersible pumps are vertical, close-coupled, heavy duty centrifugal pumps designed to work submerged in the water which they are pumping. There is no suction piping, and the pump impeller is at the very bottom of the unit. Both motor and pump are submerged. The motor is watertight, and double sealed from the surrounding water by oil or air. Better quality motors are equipped to signal an alarm when a problem senses leakage in the outer seal. Wastewater submersibles are designed to be non-clogging, and should easily pass 3 inch solids. A special design is the grinder pump, which incorporates a stainless steel knife blade on the inlet side of the impeller to chop solids to a slurry for pumping through small diameter force mains. Submersible pump capacities range from 50 to 7000 gpm; heads range from 15 to 200 ft., and power requirements range from ½ to over 100 hp.

Among the advantages to using a submersible pump are space and equipment savings: There is no need for a pumphouse. Control panels at ground level above the wet well can be locked, or installed at a remote location. Ventilation, lights, dehumidifier, sump pump of the pumphouse are not needed. This adds aesthetic value, as well as construction and operating cost savings to the installation. Submersible pumps are quiet, operating below ground and underwater. There is no danger of the motor flooding. Guide rail system enables easy servicing. With no suction piping, and the pump on only when water pressure (depth) is sufficient, problems caused by lack of suction head are eliminated. The surrounding water constantly cools both pump and motor, extending life and making them quite efficient.

Pressure Sewer Systems

If the terrain is rocky, the water table extremely high, or it is necessary to minimize sewer detention time, a gravity flow wastewater collection system may not be practical; a pressurized sewer system may be the best choice. Sewage is pumped directly from each individual house into one or more trunk lines which lead to the treatment plant. There are two main types:

Grinder Pump System: each home is equipped with a wet well (about 50 gallons) and a submersible grinder pump. Wastewater flows from the home into it; from there it is pumped into the system and to the treatment plant.

STEP System: (Septic Tank Effluent Pump). This system was designed to make use of already existing septic tanks. Water flows from the home to the septic system, and the effluent runs into a small holding tank instead of a drainfield. From there it is pumped by a non-clogging submersible pump into

the collection system. This eliminates the need for the grinder pump, with its extra maintenance.

Pressure sewer systems have the advantage of eliminating most of the groundwater infiltration encountered in gravity systems. Because the small diameter PVC piping is buried just below the frost line, installation cost can be lower than a gravity system. Disadvantages are in added operation and maintenance costs incurred due to the need for mechanical equipment at each point of entry to the system. To estimate flow in sizing pumps:

Numbers of
homes to be serviced x 3.5 people/home x 100 gpd/cap. = gpd to be pumped

Deep Well Pumps

The typical deep well pump is a vertical turbine pump. This is a centrifugal pump resting on a solid foundation with a shaft extending all the way down the well. Multi-stage impellers are arranged near the bottom of the shaft. There is no suction piping; the well water level must be over the lowest pump bowl. The well must be straight and vertical, and correct alignment of shaft sections and column pipe is very important.

A well pump follows its characteristic curve like any other centrifugal pump. Capacity is dependent on head, and well drawdown is part of that head. The pump curve applies to each impeller separately. Drawdown should remain constant while the pump is in service. Weather may have a significant effect on well pumping; dry years lower the water table; pump head increases, flow decreases. Submersible pumps with multi-stage impellers are also in use as deep well pumps and are ideally suited to crooked wells.

Jet Pumps

These are inexpensive centrifugal pumps which provide extra suction lift for the operation of small wells which must draw water up from greater than 20 ft. below the pump. Installed on pump discharge in small ones, and at well bottom on larger pumps, is a venturi apparatus which combines a recirculating flow with the incoming water; the Velocity Head is increased enough to lift the water up to the pump.

Well Pumps - Plotting Field Efficiency

The efficiency of a pumping system in operation should be checked yearly if the pump is running well. The pump curve shows the conditions of head and flow that will yield maximum efficiency. Anything off this will be less efficient, but operating conditions should still fall on the curve.

Read the metered gpm. Refer to pump curve for the head. Now measure the head: measure distance to pumping water level, from ground. Add this to the lift above ground. Add the head losses. This is the <u>Total Head</u>. This applies to the entire pump. Divide it by the number of impellers. Now it should match the head read on the curve. If it doesn't, the headloss calculation is incorrect. There may be more headloss than recognized because pump is old, clogged, C factor is off, etc. Doing this helps to identify problems in the system. This is the <u>operating point</u>. Check pump efficiency. To obtain motor horse power, read the electric meter on the pump motor. Being an accumulator, it will be ticking off Kilowatts. Time the number of Kilowatts per minute. Multiply by 60 to obtain Kw/hr. Divide by .746 to change it to Hp. This is Motor Hp, the amount of electricity actually drawn to run this pump system. Now refer to Hp for the operating point of the pump curve (bhp). Multiply it by the pump efficiency (or calculate gpm x hd/3960) to obtain water horsepower. This is the work the pump is actually doing. Divide whp by mhp; this is <u>wire to water efficiency</u>, the <u>FIELD EFFICIENCY</u>.

pump appurtenances: the pressure gage

A pressure gage on a pump should be installed on the discharge piping as far from the pump as practical. If mounted on the pump casing, it will not read accurately, as the energy in the liquid has not yet been fully converted to pressure. A properly operating gage can be a handy tool to diagnose pump problems. If readings begin to show a steady increase in pressure, there may be an obstruction building downstream from the gage. If gage readings show a steady decrease in pressure, check the possibility of a leak or break in the pipe downstream. If this turns out to be the case, turn off the pump and read the gage pressure. The reading will show the elevation of the break.

A discharge gage can be used to check the actual shutoff head. Close the discharge valve, and leave the pump running. The gage will read the true shutoff head. If this value is lower than the shutoff head on the pump curve, there may be internal pump wear or damage.

A discharge gage will yield the approximate flow rate. Read the gage, and refer to pump curve for the flow. (If there is suction head, subtract from gage reading if there is suction lift, add to gage reading).

pump operation guidelines

Selection: In selection of any pump, know the exact nature of the liquid to be pumped (specific gravity, vapor pressure, entrained air, temperature, corrosive properties, type and concentration of solids). In wastewater applications, solids are most important; at a point they change the specific gravity, and it will require more horsepower to do the same amount of work. For abrasive solids, special pump linings will be needed. Stringy materials may require special impellers. Advise the manufacturer of the water characteristics, the capacity needed and the type of drive desired.

Analyze suction and discharge conditions. Know whether service will be continuous or intermittent, what type of drive will be used, and what power supply is available.

Be aware of location conditions (space, elevation, population density), capital and operating costs, and transportation limitation.

Select the pump which will yield the required capacity under the worst operating conditions.

Installation: Install the pump in light, dry, clean location, and as low as possible. The foundation should be rigid, the driver properly aligned. Appurtenances should include vent valves at high points on the line, drain valves, relief valves, bypass connections, gages, flowmeter, and source of seal water. Suction piping should be of large diameter and minimum length, and should always be larger than inlet port of pump. Make sure spare parts and standby equipment will be available.

Operation: Follow instruction manual on startup and shutdown. Operate the equipment within proper range of flows, pressures, temperatures. Do not throttle suction to reduce pump capacity. Do not use excessive lubricant or cooling water. Do not run it if it is noisy. Avoid sudden temperature changes. Run spare equipment occasionally. Set up scheduled maintenance and inspections. Run the pump at highest efficiency possible to prevent bearing failure. A pump handles liquids. Keep air out!

Maintenance: Perform scheduled maintenance and inspections. Do not open equipment unless diagnosis indicates need; follow manufacturer's instructions. Clean internal surfaces and repaint where indicated; replace gaskets; examine for corrosion, wear, etc.

The weakest part of the pump is the stuffing box. Restore packing areas to proper condition. Be careful mounting bearings. Keep a complete record of inspection and repairs.

PROBLEMS

1. A pump has an impeller diameter of 14 inches and operates at 1750 RPM.
   A. What is the velocity of the water circulating at the impeller rim?
   B. What is the pressure developed by this pump?

2. A centrifugal pump delivers 300 gpm when its impeller speed is 1750 RPM and the total dynamic head is 57 psi. Determine:
   A. The pump's capacity if speed is reduced to 1250 RPM.
   B. The developed head in the pump if the speed is reduced to 1250 RPM. Assume ideal pump operation.

3. A pump produces 90 ft. of head and 500 gpm with a 7.5 inch diameter impeller. After several years, the impeller wears to a 7.25 inch diameter. What are the new head and flow for the worn pump?

4. With the pump whose characteristics are represented by curve #2, operating at 1800 gpm, find the change in head, flow, and horsepower when a 6 inch impeller is replaced by a 6½ inch impeller.

5. A industrial process pump is taking 400 gpm from a tank through a 6 inch diameter line from a submerged bottom inlet. The water depth over the inlet is 15 ft; the distance from the tank to the pump is 35 ft; the pump elevation is 8 ft. above inlet. What is the pressure reading at the suction inlet of the pump?

6. NPSHR for a particular pump is 6 ft. Installed into the system described below, determine the NPSHA, and the probability of efficient operation of the pump.
   Suction lift 15 ft.
   Suction head loss 12 ft.
   Flow 300 gpm
   Suction side inlet 4 inches
   Vapor pressure 2 ft.

7. A 3000 gpm 12 inch diameter turbine well pump is located 4000 ft. above sea level, and is pumping water at a maximum temperature of 90 degrees F. Suction head losses are 2 ft; water level is never less than 8 ft. above the first stage impeller. What is the NPSHA?

8. NPSHR is 7 ft. Flow is 800 gpm through 200 ft. of 8 inch diameter suction pipe. The lift is 15 ft. on the suction side. The pump is pumping cold water at sea level.
   A. What is the NPSHA?
   B. Is it adequate to lift the water without cavitation?

9. A new pumping installation is being designed: Water at 110 degrees F is to be pumped at a rate of 55 gpm from a holding tank which is open to atmosphere. The gage pressure at the end of the discharge line must be at least 56 psig. The discharge end is 30 ft., and the pump suction is 6 ft. above the water level in the holding tank. The discharge line will consist of 450 ft. of 2 inch diameter new steel pipe. Suction pipe is 10 ft. long, 2 inch diameter. There are 4 90 degree standard elbows and 1 open globe valve in the discharge line. The pump efficiency is 83%. Determine the following:
    A. Head which the pump must develop.
    B. Horsepower input
    C. NPSHA

10. The center of the intake end of a suction pipe is 5 ft. below the water surface in the river. The pipe has a uniform rise of 1 ft. per 100 ft. to the pump. Friction loss at pipe entrance is 5 ft; friction loss in the pipe is 1 ft. per 1000 ft. What is the greatest length the pipe can have without causing the pressure to drop more than 6 psi below atmospheric pressure?

11. Referring to pump shown on curve #2 (Chapter 13)
    Flow = 1600 gpm
    Suction Head 25 ft.
    Discharge reservoir elevation 100 ft.
    Discharge line 2500 ft. of 10 inch diameter 10 yr. old CIP
    Timed .85 Kw./min. on electric meter
    What is the field efficiency?

CHAPTER 14

POSITIVE DISPLACEMENT PUMPS

Whereas centrifugal pumps are meant for low pressure, high flow applications, positive displacement pumps can achieve greater pressures, but are slower moving, low flow pumps. They are used for pumping sludges and slurries and other high density or viscous fluids for which other types of pumps are inappropriate. Tolerances are close between stationary and moving parts, and flow is constant. A given volume of liquid is positively displaced with each pump cycle; there is no backwards slippage of fluid, and high shutoff pressures are developed. When pumping head increases from a blockage, or when system head increases, the pump will still move the same amount of fluid, but will draw more horsepower, and create high discharge pressure in overcoming that head.

This is quite different from the operation of the centrifugal pump, which delivers less flow and uses less horsepower under high head conditions. With a positive displacement pump, total blockage of pump or line, whether caused by foreign material or a closed valve, will cause failure of pump, line or motor.

> **Note:** Never start a positive displacement pump with the discharge valve closed!

These pumps operate at much slower speeds than centrifugal pumps. They are much larger, heavier, and more costly. The large number of moving parts (valves, cams, reciprocating components) make them more complex, and the close running clearances make them more subject to wear, with resulting parts replacements necessary. Efficiency is lower and maintenance cost higher than in the centrifugal pump.

There are two basic categories of positive displacement pump: the reciprocating pumps, and the rotary pumps. Under reciprocating pumps are generally included the plunger, or piston pump, the diaphragm pump, and the peristaltic pump. Under rotary pumps are included the rotary gear pump, the vane pump, the progressive cavity pump, and the Archimedes screw.

reciprocating pumps

These pumps have suction and discharge check valves, and the flow is cyclical; there is a to and fro motion; flow pulsates.

Plunger Pumps

Often seen in wastewater applications as a sludge pump, this reciprocating pump uses a single piston to force liquid from the suction side to the discharge side of the pump. The piston is located in the middle of the length of horizontal large diameter pipe; there are check valves on either side. As the piston rises, the suction check valve opens to allow liquid to enter the piston chamber, and the discharge check closes to prevent previously discharged liquid from backing up into the pump. As the piston lowers, the suction check closes and the discharge check opens, allowing forward movement of the liquid. Clearance between the piston and chamber walls is filled by lubricated packing, in a packing gland. The piston is connected to the driver by a connecting rod, and there are pressure gages on suction and discharge side of the pump whose readings are a good indication of what is going on inside the pump. Pressure relief devices are usually associated with this type of pump, for use when blockages occur and pressure builds up; often they are connected to an alarm. On sludge pumps, the check valves are balls made of hard material; both valve seat and balls will wear and are replaceable. The presence of the check valves make this type of pump essentially self-priming.

Figure 14.1 - Wastewater primary sludge pump.

Another type of piston pump is referred to as the lift pump. This is the typical backyard hand operated pump; a four stroke cycle is necessary to operate it. A bucket valve is built into the piston and moves up and down with it.

A. Air exhaust - piston descends to bottom of cylinder, forcing out air.

B. Water inlet - upward stroke - vacuum is created. Water flows into cylinder.

C. Water transfer - downward stroke - water flows through the bucket valve, and is transferred to upper side of piston.

D. Water discharge - piston rises - water is discharged from pump.

A piston pump which is a deep well pump has the piston at the bottom of the well. Like centrifugal pumps, positive displacement pumps with a suction lift generally are restricted to about twenty feet of practical lift.

Other types of piston pumps are used in industrial applications, often with spring actuated multiple pistons which move rapidly. These are in use for their ability to develop very high pressures, and are not meant for solids handling.

Many large pumping stations equipped with piston pumps were built during the nineteenth century. Enormous engines were employed. The development of the electric motor and centrifugal pump provided a compact pumping unit which has virtually eliminated these pumps with their pulsating flow from water service.

Diaphragm Pumps

In water utilities, diaphragm pumps are most often used for chemical feed. The pumping chamber is in a vertical position with the check valves above and below the diaphragm chamber. The principle of operation is the same as that of the plunger pump. The pumping mechanism is controlled by a rod which moves back and forth on a cam, and pushes the diaphragm in and out, moving the

liquid. Percent stroke of the diaphragm is controllable by setting, and is the means of controlling flow. Checks are sometimes shaped as discs, rather then balls, and are much lighter weight than those on plunger sludge pumps. If the pump is located above the liquid to be pumped, priming may be necessary. Diaphragms and checks will wear easily, and are replaceable. There are several variations of the diaphragm pump available and in use, referred to as chemical metering pumps.

Figure 14.2 - Diaphragm pumps in chemical feed room.

Peristaltic Pumps

Peristaltic pumps are not true reciprocating pumps, as they do not have a back and forth action, but they do however, produce a pulsating flow, and so are included here. These are mainly used as chemical feed pumps also, usually for very small flows, and are controlled by a set of fingers, or a moving ball, which repeatedly squeezes flat a length of tubing filled with the liquid to be pumped, expelling it. As the tubing expands again, it fills with incoming liquid, which waits to be squeezed forward again. There are no valves associated with this type of pump. Movement is strictly by peristaltic action. Tubing is replaced as necessary.

rotary pumps

Rotary pumps are positive displacement pumps which are primarily used as a source of fluid power in hydraulic systems, for auto, aircraft, and mobile equipment applications. A few types are in common use in water utilities.

Progressive Cavity Pumps

Chemical slurries, polymers, and sludges are often moved with progressive cavity pumps, a type of rotary pump which can achieve pressures up to 2000 psi, depending on pump length. The liquid travels in the spaces between a rigid rotor (hardened

Figure 14.3 - Progressive Cavity pumps used for moving sludge at wastewater plant.

chrome steel) and a flexible stator (rubber). It operates in a manner something like a snake swallowing a frog; as each section closes, it opens the section in front of it for the liquid to move into: a cavity progression. The rotor revolves rapidly, capacity is directly proportional to rotor speed and pump size, and a non pulsing flow is produced. There are no valves associated with the pumping, but packing is needed to prevent liquid from leaking out of the pump and up the shaft. Rotor and stator are always in contact as they move, and the liquid pumped does the lubricating. This type of pump should never be run dry, as it will quickly burn up the stator and fail.

Rotary Impeller Pumps
Used for oil transfer and hydraulic services, the "impeller" is sometimes soft and flexible, and scoops the liquid in and out of the pumping chamber. Its appearance is that of a centrifugal pump, but the impeller speed is lower, and the close tolerance between impeller and casing make it positive displacement.

Rotary Gear Pumps - intermeshing gears, or lobes, trap the fluid between them as they rotate. Gear pumps can withstand rugged operating conditions, and are simple in construction, and economical in cost and maintenance. They can handle many types of liquids (including air/liquid mixtures) over a wide range of capacities and pressures, but are best adapted to low pressures and flows under 500 gpm. These are sometimes called Hydraulic

pumps, and work well with suction lifts. Because of the constant close meshing of the gear lobes, they are best adapted to liquids having lubricating qualities.

Rotary Vane Pumps - used for high pressure industrial applications, often with hot liquids. The vanes slide in and out as the pump chamber fills and empties. Vane pumps emphasize quiet operation, long life, and little load on the bearings.

Screw Pumps

The Archimedes Screw was invented by the ancient Greek himself about 287 BC (perhaps the world's first great physicist). This is a large piece of lifting apparatus, not truly a pump at all, but in fact a screw with deeply cut threads, laid at an angle up to 45 degrees, with its lower end submerged in the water. The liquid submergence at entrance, and the speed of revolution control the output. As the pump revolves, the water "crawls"

up the screw. Maximum lift is about 25 ft., but the pump can be built in stages for higher lifting. An angle of more than 45 degrees of incline results in backwards slippage of the water, and is inefficient. The screw pump is relatively maintenance free, passes large solids, and is economical to operate. Limitations are with maximum lift, and with size. This type of pump is often seen as a raw wastewater pump at headworks of small treatment facilities. Laid horizontally, screw conveyors are in use for moving grit, dewatered sludges, and other high solids materials.

Figure 14.4 - Screw pump used to pump wastewater influent.

miscellaneous pumping apparatus

Pneumatic Ejector: The pneumatic ejector is based on reciprocating operation, and is a fill and draw affair. The lift it achieves is based on air pressure, not water pressure. In use in small wastewater lift stations, this type of pumping apparatus is excellent for water containing solids. There are no rotating parts to clog, and it can handle flows up to 300 gpm. The water flows by gravity into a small vented chamber. When full, the vent closes, compressed air is introduced, and the water is forced out into the discharge line.

The vent opens, and new water comes in. The cycle lasts about a minute, and the only machinery is an air compressor.

Air Lift Pump: The air lift pump, also known as the "bubble pump", is also used in small wastewater lift stations, and for pumping activated sludge; its advantage over the pneumatic ejector is that a smooth constant flow is produced. The chamber includes a vertically positioned discharge pipe which extends to the bottom of the tank. Compressed air is fed in and saturates the water in this pipe, decreasing its density. As it moves up and out, new water fills in behind it, is filled with air and rises, thus producing a constant discharge. As with the pneumatic ejector, the only power needed is to run the air compressor, and there are no moving parts. Lift depends on the differential in hydrostatic pressure, and is limited; best operation is achieved with horizontal flow.

Vacuum Pumps: Vacuum is any pressure less than atmospheric. Vacuum pumps can be rotary or piston pumps, completely oil sealed, which just pump air out of a chamber, creating vacuum.

They can also be designed as a centrifugal pump, called a Liquid Jet Vacuum Pump. These have a special attachment called an ejector, or eductor. This is a venturi constriction with a side outlet; as velocity increases passing through the venturi, the side outlet, open to atmosphere draws air in, creating a constant useful vacuum as long as the pump operates. The air/water mixture is then discharged into a tank where air and water separate, and only water is recycled through the pump. In water treatment facilities, the eductor is used to move slurries (lime feed, filter sand), to provide gas flow (injector of vacuum chlorinator), and to create vacuum in the laboratory (at sink faucets).

Blowers: Blowers pump air. Used for wastewater aeration service, they are large in size, and power demanding. Design may be positive displacement (rotary gear) or centrifugal (turbine). The centrifugal types run at higher speeds, but are less efficient than the rotary blowers. Drawing in ambient air through a filter, they transport it at low pressures to the underwater point of application through a large diameter discharge pipe called a header. Near the end of the discharge line, the header splits up into multiple diffusers for application of air to the water. All blowers are noisy.

Figure 14.5 - Wastewater aeration tank blower.

Hydraulic Ram: The ram lifts water by utilizing the impulse developed when a moving mass of water is suddenly stopped. In other words, it makes use of the pressure created by water hammer. Not an efficient pumping mechanism, a relatively large amount of water must be available at moderate head in order to move a small amount of water to higher head. Water

flows from a reservoir to a valve box below, and on out of the box through a waste valve. If that valve is suddenly closed, water hammer pressures develop, which open another valve into a vertical delivery pipe, and some water passes up the pipe. As negative pressure waves return, the delivery pipe valves closes, and water continues on through the box. What keeps it going is the water flow through the waste valve. When well established, it provides enough force to shut the waste valve, starting a new cycle. The hydraulic ram is not really a pump - just a set of valves.

Turbine Generator (The Backwards Pump): A motor converts electrical energy to mechanical energy to turn a shaft. This, connected to a pump shaft, will rotate the pump impeller. The system expends electrical energy to produce water energy (pressure).

The opposite principle, that of the turbine generator, uses water energy to turn a turbine (impeller) connected to a motor shaft, to drive a motor, thus creating electrical energy. In the standard mode, the essential water energy would be provided by a waterfall - on the turbine, to keep it moving. But water pressure is sometimes wasted in distribution systems, and can be captured for this type of use. For instance, a large pressure reducing valve between a water supply and a city's water system reduces and controls pressure so that it remains at a useable level. The excess is wasted by the reducing valve. If a pump were installed there (a turbine with a motor), adapted to run backwards, the excess pressure could be captured and converted to electricity. The water system would still benefit from the operation of the pressure reducing valve, and the electricity created by the adaptation could be used to run other equipment. An ordinary pump used this way would need modifications; too much flow would run the impeller too fast for efficient or safe transmission of power to the motor; damage may result to mechanical parts, and control valves are needed. The backwards pump principle is now being put to use in installations across the country.

Hydropneumatic Storage: When stored water is desired, but elevated storage is not practical, hydropneumatic storage may be the best choice for small community water systems, private wells, industries, apartment buildings - anyplace where capital funds and/or space is limited. In this type of installation, air pressure is in use to equalize water pressure, prevent surges, and minimize pump operation time.

A relatively small tank is employed, which has an air chamber, and a water chamber, with a bladder or diaphragm separating the two. Pressure sensors are set to a range of pressures at the tank. When pressure in the system is low, the pump operates, filling the water chamber. The diaphragm flexes toward the air chamber, compressing the fixed amount of air in that section.
This increases air pressure - to the maximum setpoint, and the pump switches off. As the water is used, the tank empties, the diaphragm relaxes, and pressure decreases to the low setpoint; then the pump switches on again. This allows system pressure to vary over a range of values, decreases operating time for the pump, and the accompanying pressure transients which are associated with a constant on/off mode.

The result is the same as filling and emptying an elevated tank, but air pressure is replacing the water pressure that elevation would provide. Capital cost for hydropneumatic storage is less than elevated storage, but operating costs are greater.

PROBLEMS

1. A piston pump chamber is 12 inches in diameter and 18 inches deep. How many gallons of sludge are moved with each stroke?

2. A pump chamber is 10 inches in diameter and 12 inches deep. Pump operates at 1 stroke per second. What is the gpd pumping rate?

3. Primary sludge is pumped to the digester 4 times per day, for a 15 minute pumping cycle each time. If the piston chamber is 12 inches in diameter and 16 inches in depth, pumps at 1 stroke per second, and is set at 3/4 stroke, what is the gpd pumping rate?

4. Primary sludge is pumped to an empty digester, which is 50 ft. in diameter and 30 ft. deep. The pump cylinder is 6 inches in diameter, 8 inches in depth, is set at full stroke, and pumps 1 stroke every two seconds. Sludge is pumped 4 times per day, for a 10 minute pumping cycle each time. At the end of a 2 month pumping period, how deep is the sludge in the digester?

5. A plunger pump with a bore of 5 and 3/16 inches, and a stroke of 2 and 5/32 inches is operating at a rate of 65 revolutions per minute.
   A. What is the gpm delivery?
   B. How many gallons of liquid would be pumped in a 30 day month if the pump operates for 4 hours every day?

6. A piston pump with a 6 inch diameter piston and a 1 ft. depth chamber is set on ½ stroke, at a pumping rate of 500 strokes per minute. It pumps water out of a cylindrical tank 40 ft. in diameter and 50 ft. depth.
   A. What is the drawdown in 4 hours?
   B. If water were also flowing into this tank at 1 cfs, while the pump is running - is the tank filling or emptying, and by how much?

7. Given two piston pumps. Each piston is 8 inches in diameter, and has a stroke of 8 inches. The rate is 50 stroke per minute. They are pumping liquid out of a circular tank 24 ft. in diameter, with a 50 ft. liquid depth. The drawdown after 45 minutes of pumping is 2 ft.
   A. What percent of their capacity are the pumps pumping?
   B. If pumping less than full capacity, what mechanical problem would this most likely indicate?

APPENDIX I

PROBLEM SOLUTIONS

chapter 1

1. **Answer**   98.3 gallons

   Water weighs 8.34 lb./gal. Therefore:

   $$\frac{820 \text{ lb.}}{8.34 \text{ lb./gal.}} = 98.3 \text{ gal.}$$

2. **Answer**   8340 lb.

   Wt. Water = 8.34 lb./gal. x 1000 gal. delivered for a dollar = 8340 lb.

3. **Answer**   A. .7
                 B. 5.84 lb./gal.

   A. Density gasoline = $\frac{131 \text{ lb.}}{3 \text{ cu.ft.}}$ = 43.7 lb./cu.ft.

   $$SG = \frac{\text{density gasoline}}{\text{density water}} = \frac{43.7 \text{ lb./cu.ft.}}{62.4 \text{ lb./cu.ft.}} = .7$$

   B. Multiply Specific Gravity by weight of a gallon of water.

   8.34 lb./gal x .7 = 5.84 lb./gal.

4. **Answer**   36.7 lb.

   Solve for specific gravity; get lb./gal; multiply by 5.

   $$SG = \frac{\text{wt. oil}}{\text{wt. water}} = \frac{55 \text{ lb./cu.ft.}}{62.4 \text{ lb./cu.ft.}} = .88$$

$$SG = \frac{\text{wt. oil}}{\text{wt. water}}$$

$$.88 = \frac{\text{wt. oil}}{8.34 \text{ lb./gal.}}$$

7.34 lb./gal. = wt. oil

7.34 lb./gal. x 5 gal. = 36.7 lb.

5. **Answer** 79.25 lb./cu.ft.

62.4 lb./cu.ft. x 1.27 = 79.25 lb./cu.ft.

6. **Answer** A. 41,700 lb. water
B. 76,311 lb. sulfuric acid

A. Multiply by density of water.

5000 gal. x 8.34 lb./gal = 41,700 lb. water

B. Solve for density of one gallon sulfuric acid; multiply by 5000 gal.

$$SG = \frac{\text{density sulfuric acid}}{\text{density water}}$$

$$1.83 = \frac{\text{density sulfuric acid}}{8.34 \text{ lb./gal.}}$$

15.26 lb./gal. = density sulfuric acid

15.26 lb./gal x 5000 gal. = 76311 lb. sulfuric acid.

7. **Answer** 2.4

$$SG = \frac{\text{density concrete}}{\text{density water}} = \frac{150 \text{ lb./cu.ft.}}{62.4 \text{ lb./cu.ft.}} = 2.4$$

8. **Answer** A. 848.64 lb./cu.ft.
B. 22914 lb./cu.yd.

A. Multiply SG mercury by density water.

13.6 x 62.4 lb./cu.ft. = 848.64 lb./cu.ft.

B. Multiply density mercury by number cu.ft. per cu.yd.

848.64 lb./cu.ft. x 27 cu.ft./cu.yd. = 22914 lb./cu.yd.

9. **Answer**   2700 lb.

   Multiply by number of cu.ft. per cu.yd.

   100 lb./cu.ft. x 27 cu.ft./cu.yd. = 2700 lb.

---

10. **Answer**   A. .1875 lb./cu.ft.
                   B. 180 lb.

   A. Multiply by density of air (.075 lb./cu.ft.).

   .075 lb./cu.ft. x 2.5 = .1875 lb./cu.ft.

   B. Multiply density of chlorine by cu.ft. volume of room.

   Vol. = 10 ft. x 12 ft. x 8 ft.

   Vol. = 960 cu.ft.

   960 cu.ft. x .1875 lb./cu.ft. = 180 lb. chlorine in room

---

11. **Answer**   A. 47936 lb./day
                   B. 1.33

   A. Solve for lb. alum/gal.; multiply by gal./min. = lb./min.; multiply by 1440 min./day = lb./day.

   $$\frac{83 \text{ lb./cu.ft.}}{7.48 \text{ gal./cu.ft.}} = 11.1 \text{ lb./gal.}$$

   11.1 lb./gal x 3 gal/min. pumped = 33.3 lb./min. pumped

   33.3 lb./min. x 1440 min./day = 47936 lb./day

   B. $SG = \dfrac{\text{density alum}}{\text{density water}} = \dfrac{83 \text{ lb./cu.ft.}}{62.4 \text{ lb./cu.ft.}} = 1.33$

---

12. **Answer**   A. 116532 lb.
                   B. 3.7 hrs.

   A. Solve for bed volume (cu.ft.); solve for lb./cu.ft. sludge; multiply.

   Vol. bed = 30 x 60 x .83 = 1494 cu.ft.

   $SG = \dfrac{\text{wt. sludge}}{\text{wt. water}}$

   $1.25 = \dfrac{\text{wt. sludge}}{62.4 \text{ lb./cu.ft.}}$

   78 lb./cu.ft. = wt. sludge

78 lb./cu.ft. x 1494 cu.ft. = 116532 lb. sludge applied to bed

B. Convert vol. bed to gallons; divide by 50 gpm.

1494 cu.ft. x 7.48 gal./cu.ft. = 11175 gal. needed to fill bed

$\frac{11175}{50 \text{ gpm}}$ = 223.5 min. = 3.7 hrs.

---

13. **Answer** It will sink all the way to the bottom.

---

14. **Answer** ½ inch

Volume is given; solve for depth. Weight has no significance.

1 gal. = .134 cu. ft.

Vol = .785 d^2 x depth

.134 = .785 x 2^2 x depth

.043 ft. = depth

½ inch = depth

---

15. **Answer** 11 inches

New water depth = normal depth - grit depth. Solve for volume of grit.

62.4 lb./cu.ft. x 2.2 SG grit = 137.3 lb./cu.ft. density grit

$\frac{22000 \text{ lb. grit}}{137.3 \text{ lb./cu.ft. density grit}}$ = 160 cu.ft. volume of grit

Vol = l x w x h

160 = 30 x 5 x h

1.1 ft. = h

13 in. = h

24 in. water - 13 in. grit = 11 in. water

---

16. **Answer** A. .64 cu.ft.
B. 2.25

A. The stone displaces 40 lb. of water: solve for cu.ft. of water displaced (equivalent to volume of stone).

$\frac{40 \text{ lb. water displaced}}{62.4 \text{ lb./cu.ft.}}$ = .64 cu.ft. water displaced

B. Stone volume = .64 cu.ft.; this weighs 90 lb.

$$\frac{90 \text{ lb.}}{.64 \text{ cu.ft.}} = 140.6 \text{ lb./cu.ft. density of stone}$$

$$SG = \frac{\text{density stone}}{\text{density water}} = \frac{140 \text{ lb./cu.ft.}}{62.4 \text{ lb./cu.ft.}} = 2.25$$

---

17. **Answer** 2.19 cu.ft.

This person will displace his total volume since his Specific Gravity is greater than that of water.

Solve for his density; then solve for the volume of person.

$$SG = \frac{\text{density person}}{\text{density water}}$$

$$1.1 = \frac{\text{density person}}{62.4 \text{ lb./cu.ft.}}$$

68.64 lb./cu.ft. = density person

$$\frac{150 \text{ lb. person}}{68.64 \text{ lb./cu.ft. density person}} = 2.19 \text{ cu.ft. volume of person}$$

This person will displace 2.19 cu.ft. of water.

---

18. **Answer** 376 gallons

Solve for volume of coiled pipe, and volume of tank. Subtract pipe volume from tank volume. Change to gallons.

Pipe
Vol. = .785 d^2 x L
Vol. = .785 x $.33^2$ x 160
Vol. = 13.7 cu.ft.

Tank
Vol. = l x w x h
Vol. = 4 x 4 x 4
Vol. = 64 cu.ft.

64 cu.ft. - 13.7 cu.ft. = 50.3 cu.ft.

50.3 cu.ft. x 7.48 gal./cu.ft. = 376 gallons needed to fill tank.

---

19. **Answer** 49920 lb.

To submerge this box, it must become the same weight as an equivalent volume of water. Solve for that weight.

Vol. = l x w x h
Vol. = 10 x 10 x 8
Vol. = 800 cu.ft.

800 cu.ft. x 62.4 lb/cu.ft. = 49,920 lb.

---

20. **Answer**   10.7 inches

Solve for volume of concrete block. This is the volume of displaced water. Apply to tank area to get depth of displacement.

Concrete Vol. = l x w x h
Concrete Vol. = 4 x 4 x 4
Concrete Vol. = 64 cu.ft.

Displacement Vol. = l x w x h
64 = 6 x 12 x h
.889 ft. = h
10.7 in. = h

---

21. **Answer**   240 lb.

If this log had the same weight as water, it would rest just barely submerged. Find the difference between its weight and that of the same volume of water. That is the weight needed to keep it down.

62.4 lb./cu.ft. (water)
- 40.0 lb./cu.ft. (wood)
22.4 lb./cu.ft. difference

Vol. wood = .785 x 1.167^2 x 10 ft. = 10.69 cu.ft.

22.4 lb./cu.ft. x 10.7 cu.ft. = 240 lb. needed to hold it below surface

---

22. **Answer**   410,400 lb./day

Solve for lb. pumped/min.; change to lb./day.

8.34 lb./gal. water x 1.14 SG liquid = 9.5 lb./gal. liquid

30 gal/min x 9.5 lb./gal = 285 lb./min.

285 lb./min. x 1440 min./day = 410,400 lb./day

---

23. **Answer**   3.7%

Solve for volume of sample; solve for volume of mudballs; solve for % mudballs.

Volume sampler = .785 x $.5^2$ x 1

Vol. = .19625 cu.ft.

Mudball volume = volume rise of water level in graduate

Volume rise = .785 x .167² x .33

Vol. = .00726 cu. ft.

$$\% \text{ mudballs} = \frac{.00726 \text{ cu.ft.}}{.19625 \text{ cu.ft.}} = .037 = 3.7\%$$

Percent mudballs in sampler = Percent mudballs in filter

---

24. **Answer**   32.5 gal.

The probe, even though suspended, registers weight on the scale - the weight of an equivalent volume of water.

Subtract weight of cylinder; solve for volume of immersed portion of probe; obtain weight of equivalent volume of water; subtract from total weight; change to gallons.

```
  300 lb. weight registered on scale
-  25 lb. weight of cylinder
  275 lb. weight of water plus immersed probe
```

Volume probe = .785 x .167² x 3

Vol. = .067 cu. ft.

Wt. of equiv. vol. of water = 62.4 lb./cu.ft. x .067 cu.ft. = 4.12 lb.

```
  275 lb. weight of water plus probe
-   4.12 lb. probe
  270.9 lb. weight of water
```

$$\frac{270.9 \text{ lb.}}{8.34 \text{ lb./gal.}} = 32.5 \text{ gal. water}$$

---

25. **Answer**   7.5 inches

Draw Diagram. Weight of concrete block = underwater weight minus upward lift from attached float. Solve for underwater weight of concrete. Solve for upward lift from float. Subtract both of these from weight of concrete in air to obtain additional lift needed to get block to rise off bottom. Obtain upward lift for each inch of submersion depth of float; divide into total needed lift to get inches rise needed to lift block off bottom.

$$\frac{1500 \text{ lb. weight of concrete in air}}{150 \text{ lb./cu.ft. density concrete}} = 10 \text{ cu.ft. of concrete block}$$

62.4 lb./cu.ft. weight of water x 10 cu.ft. of
water = 624 lb. buoyant force.

1500 lb. wt. concrete in air
- 624 lb. buoyant force
876 lb. wt. of concrete underwater

Lift from float:

Volume float = 4 x 4 x 4 = 64 cu.ft.

$\dfrac{400\text{ lb. wt. of float}}{64\text{ cu.ft.}}$ = 6.25 lb./cu.ft.

$\dfrac{6.25\text{ lb./cu.ft. float wt.}}{62.4\text{ lb./cu.ft.}}$ = 10%

10% of this float (4.8 inches) would be submerged even if it were floating unrestrained.

Therefore, only 4.2 inches of float submergence is providing lift for the concrete block.

Vol. = 4 x 4 x .35

Vol. = 5.6 cu.ft. submerged float

5.6 cu.ft. x 6.25 lb./cu.ft. = 35 lb.

5.6 cu.ft. x 62.4 lb./cu.ft. = 349.4 lb.

349.4 lb. wt. of water equiv. to lifting part of the float
-    35 lb. wt. of the lifting part of the float
314.4 lb. lift force provided by float

Therefore:

876 lb. wt. of concrete underwater
- 314.4 lb. lift from float
561.6 lb. actual weight of concrete block attached to float

$\dfrac{314.4\text{ lb. total float lift}}{4.2\text{ inches submerged float}}$ = 75 lb. lift for each inch of submersion.

$\dfrac{561.6\text{ lb. lift needed}}{75\text{ lb. lift per in. water rise}}$ = 7.5 in. water rise to lift concrete off bottom.

chapter 2

1. **Answer** 4.4 ft./sec.

   Change MGD to cfs, inches to ft.; solve for velocity

   $$Q = A \ V$$
   $$1.55 = .785 \times .67^2 \ V$$
   $$4.4 = V$$
   $$4.4 \ \text{ft./sec.} = V$$

2. **Answer** .016 ft./sec.

   Change gpm to cfs, inches to ft.; solve for velocity.

   $$Q = A \ V$$
   $$.11 = .785 \times 3^2 \ V$$
   $$.016 \ \text{ft./sec.} = V$$

3. **Answer** 4.5 ft. (54 inches)

   Change MGD to cfs; solve for diameter.

   $$Q = A \ V$$
   $$62 = .785d^2 \times V$$
   $$4.44 \ \text{ft.} = d$$

   Choose the next larger pipe size (4.5 ft.).

4. **Answer** 118,333 gal./4 hrs.

   Height of flume not needed. Change inches to ft.; solve for Q; change to gal./hr., then gal./4 hr.

   $$Q = A \ V$$
   $$Q = .83 \times .67 \times 2$$
   $$Q = 1.1 \ \text{cfs}$$
   $$Q = .71 \ \text{MGD}$$
   $$Q = 710,000 \ \text{gpd}$$
   $$Q = 29583 \ \text{gal./hr.}$$
   $$Q = 118,333 \ \text{gal./4 hrs.}$$

5. **Answer**   2962 gpm

Change ft./min. to ft./sec., inches to ft.; solve for flow.

$$Q = \quad A \quad V$$
$$Q = .785 \times 1.83^2 \times 2.5$$
$$Q = 6.6 \text{ cfs}$$
$$Q = 2962 \text{ gpm}$$

---

6. **Answer**   1.9 MGD

Change gpm to cfs; add flows.

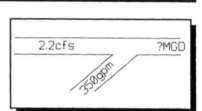

$$350 \text{ gpm} = .78 \text{ cfs}$$
$$2.2 \text{ cfs} + .78 \text{ cfs} = 3 \text{ cfs}$$
$$\frac{3 \text{ cfs}}{1.55 \text{ cfs/MGD}} = 1.9 \text{ MGD}$$

---

7. **Answer**   .257 cfs

Change inches to ft.

$$Q = A \quad V$$
$$Q = .785 \times .33^2 \times 3$$
$$Q = .257 \text{ cfs}$$

---

8. **Answer**   26.7 ft./sec.

Change velocity to ft./sec.; assuming flow is the same in both pipes (steady state), can use AV=AV and solve for velocity directly.

12 inch pipe		3 inch nozzle
A V	= A V	
$.785 \times 1^2 \times 1.67$	$= .785 \times .25^2 \times V$	
26.7 ft./sec.	= V	

---

9. **Answer**   1.3 ft.

Change MGD to cfs; convert velocity to ft./sec.; solve for depth.

$$\frac{1 \text{ mile}}{30 \text{ min.}} = \frac{5280 \text{ ft.}}{1800 \text{ sec.}} = 2.93 \text{ ft./sec.}$$

$$Q = A \ V$$

$$18.6 = 5 \text{ x depth x } 2.93$$

$$1.3 \text{ ft.} = \text{depth}$$

---

10. **Answer**   3,100,000 gallons

Solve for volume of pipe.

$$\text{Volume} = .785 \text{ x } 5^2 \text{ x } 5280 \text{ x } 4$$

$$\text{Vol.} = 414480 \text{ cu.ft.}$$

$$\text{Vol.} = 3,100,000 \text{ gal.}$$

---

11. **Answer**   A. 10771.2 gpm

$$Q = A \ V$$

$$Q = 4 \text{ x } 2 \text{ x } 3$$

$$Q = 24 \text{ cfs}$$

$$Q = 10771.2 \text{ gpm}$$

---

12. **Answer**   5,299,430 gpd

Change min. to sec.; solve for velocity.
Area of flow = ½ area of pipe.
Solve for flow.

$$\frac{350 \text{ ft.}}{150 \text{ sec.}} = 2.33 \text{ ft./sec. velocity}$$

$$Q = \qquad A \qquad V$$

$$Q = .785 \text{ x } \frac{3^2}{2} \text{ x } 2.33$$

$$Q = 8.2 \text{ cfs}$$

$$Q = 5,299,430 \text{ gpd}$$

---

13. **Answer**   .3 cfs

Change inches to ft.; change ft./min. to ft./sec.

$$Q = A \ V$$

$$Q = 2 \text{ x } .75 \text{ x } .2$$

$$Q = .3 \text{ cfs}$$

14. **Answer**   A. 2,270,000 gpd
                  B. $204,300.00/quarter

    A. Solve for flow; change to gpd.

    $$Q = A\ V$$
    $$Q = .785 \times .67^2 \times 10$$
    $$Q = 3.52\ cfs$$
    $$Q = 2,270,000\ gpd$$

    B. Multiply gpd by days/quarter; convert to units of 1000 gal.; multiply by $.

    2,270,000 gpd x 90 days/qtr. = 204,300,000 gal./qtr.

    $$\frac{204,300,000\ gal./qtr.}{1000} = 204,300 \times \$1 = \$204,300/qtr.$$

---

15. **Answer**   5.6 cfs

    Velocity changes as pipe narrows; flow remains the same.

---

16. **Answer**   .92 ft./sec.

    Velocity is dependent upon flow and cross sectional area. Length does not affect velocity.

---

17. **Answer**   .03 ft./sec.

    Assume steady state flow through pipe and clarifier. Use AV=AV; change inches to ft. and velocity to ft./sec.

Pipe		Clarifier
A x V	=	A x V
.785 x 2² x 1	=	10 x 10 x V
.03 ft./sec.	=	V

---

18. **Answer**   A. 28.7 ft./sec.
                  B. 78.9 ft./sec.
                  C. 707.8 ft./sec.

    Change inches to ft., MGD to cfs.

    A. $$Q = A\ V$$
                $$15.5 = .785 \times .83^2 \times V$$
                $$28.7\ ft./sec. = V$$

B. $Q = A \ V$

15.5 $= .785 \times .5^2 \times V$

78.9 ft./sec. $= V$

C. $Q = A \ V$

15.5 $= .785 \times .167^2 \times V$

707.8 ft./sec. $= V$

---

19. **Answer**   A. 1.5 ft.
                   B. 10.2 ft./sec.
                   C. 3 cfs

A. Steady state flow through A & B; solve for diameter.

$A \ V = A \ V$

$.785 \times d^2 \times 2.83 = .785 \times 1^2 \times 6.37$

$d = 1.5$ ft.

B. Dimensions are given; solve for velocity.

$Q = A \ V$

$2 = .785 \times .5^2 \ V$

10.2 ft./sec. $= V$

C. Solve for flow in pipe B; subtract flow in C $=$ flow in D.

$Q = A \ V$

$Q = .785 \times 1^2 \times 6.37$

$Q = 5$ cfs

5 cfs (B) - 2 cfs (C) $= 3$ cfs (D)

---

20. **Answer**   A. .22 cfs
                   B. 545 cu.meters/day

A. Convert 100 gpm to cfs.

B. Convert 100 gpm to gpd, then to L/day, then to cu.meters/day.

$$\frac{100 \text{ gal.}}{\text{min.}} \times \frac{1440 \text{ min.}}{\text{day}} \times \frac{3.785 \text{ liters}}{\text{gal.}} = 545{,}040 \text{ L/day}$$

$$\frac{545{,}040 \text{ L/day}}{1000 \text{ L/cu.meter}} = 545 \text{ cu.meters/day}$$

21. **Answer** .4 ft.

Depth is a component of cross sectional area. Flow is steady, velocity at the constriction increases because of width decrease. Assuming that downstream of constriction velocity and width return to normal, depth should remain constant.

22. **Answer** A. 12 inch section velocity = 3.75 ft./sec.
    B. 10 inch section velocity = 5.45 ft./sec.

Draw Diagram.

A. Change inches to ft.; using AV=AV, solve for vel. 12 inch section.

6 inch	12 inch
A V	= A V
$.785 \times .5^2 \times 15$	= $.785 \times 1^2 \times V$
3.75 ft./sec.	= V

B. Same method, solve for velocity in 10 inch section.

6 inch	10 inch
A V	= A V
$.785 \times .5^2 \times 15$	= $.785 \times .83^2 \times V$
5.45 ft./sec.	= V

23. **Answer** .0167 ft./sec

Change MGD to cfs and gal. to cu. ft.; using volume, calculate area of tank.

Volume tank = Area x Length

802.1 = A x 20

40.1 sq.ft. = A

Q = A   V

.67 = 40.1 x V

.0167 ft./sec. = V

24. **Answer** 23.3 ft./sec.

Draw Diagram. Change inches to ft.; solve for flow in 2 inch diameter pipe, and subtract from flow in 6 inch diameter pipe = flow in 4 inch diameter pipe. Solve for velocity in 4 inch pipe.

Q = A V

Q = .785 x .167² x 40

Q = .88 cfs

2.87 cfs (6 inch) - .88 cfs (2 inch) = 1.99 cfs (4 inch)

Q = A V

1.99 = .785 x .33² x V

23.3 ft./sec. = V

---

25. **Answer** 2.4 cfs

Draw Diagram. Change MGD to cfs, inches to ft.; solve for recirc. flow, subtract from TF flow.

Q = A V

Q = .785 x .67² x 2

Q = .7 cfs - being recirculated

Plant flow + Recirc. flow = Trickling filter flow

X + .7 cfs = 3.1 cfs

X = 2.4 cfs - incoming plant flow

---

chapter 3

1. **Answer** 242.6 ft.

Convert pressure to ft. of water.

105 psi x 2.31 ft./psi = 242.6 ft.

2. **Answer**   A. 60.6 psi
                B. 31.2 psi

Draw Diagram.

A. Gage B registers pressure from tank elevation, and from water depth.

> 2290 ft.
> - 2150 ft.
>   140 ft.

> 140 ft. div. by 2.31 ft./psi = 60.6 psi

B. With both valves closed, gage B registers only the pressure in the pipe due to elevation difference. The water in the tank does not register.

> 2222 ft.
> - 2150 ft.
>   72 ft. div. by 2.31 ft./psi = 31.2 psi

---

3. **Answer**   A. 25.6 ft.
                B. 11.1 psi

A. Change MG to cu.ft.; using volume, solve for depth.

$$\frac{1,500,000 \text{ gal.}}{7.48} = 200535 \text{ cu. ft.}$$

Vol. = .785 x d^2 x depth

200535 = .785 x 100^2 x depth

25.6 ft. = depth

B. Convert to psi.

$$\frac{25.6 \text{ ft.}}{2.31} = 11.1 \text{ psi}$$

---

4. **Answer**   A. 150.2 ft.
                B. 9361 psf

A. Convert pressure to ft. of water.

> 65 psi x 2.31 ft./psi = 150.2 ft.

B. Convert psi to psf.

> 65 psi x 144 sq.in./sq.ft. = 9361 psf

5. **Answer**   4571 psi

Change miles to ft., then to psi.

$$5280 \text{ ft./mi. x } 2 = 10560 \text{ ft.}$$

$$\frac{10560 \text{ ft.}}{2.31 \text{ ft./psi}} = 4571 \text{ psi}$$

---

6. **Answer**   2179.8 ft.

Draw Diagram.   Change psi to ft.   Gage registers height of bluff plus depth of water.

$$62 \text{ psi x } 2.31 \text{ ft./psi} = 143.2 \text{ ft.}$$

gage reading = bluff height + water depth

$$143.2 = N + 75$$

$$68.2 \text{ ft. } = N$$

2248 ft. bluff elevation
- 68.2 ft.
2179.8 ft. customer's house elevation

---

7. **Answer**   54.8 ft. of oil

The weight of oil is .84 of the weight of water.

$$\frac{46 \text{ ft. water}}{.84} = 54.8 \text{ ft. oil}$$

---

8. **Answer**   58.9 psi

Mercury is 13.6 times heavier than water.

$$.433 \text{ lb./ft./sq.in.(wt. water) x } 10 \text{ ft. x } 13.6 = 58.9 \text{ psi}$$

---

9. **Answer**   100 ft. head

Head on pipe = ft. of pressure

$$\text{Pressure } = \text{ Weight x Height}$$

$$6234 \text{ psf } = 62.4 \text{ lb./cu.ft. x height}$$

$$100 \text{ ft. } = \text{ Height}$$

10. **Answer**   A. -11.3 ft.
                 B. -4.9 psi

A. Change to inches of water; then to ft. water.

       10" mercury x 13.6 (SG mercury) = 136" water = -11.3 ft. water

B. Change to psi.

$$\frac{-11.3 \text{ ft.}}{2.31 \text{ ft./psi}} = -4.9 \text{ psi}$$

This is a vacuum condition (less than atmospheric pressure). It occurs when pumps draw a suction lift.

---

11. **Answer**   203.2 psi

Draw Diagram. Convert pipe pressure to ft.; add for elevation; change back to psi.

         200   psi x 2.31 = 462 ft.

            462   ft.
     +   7.5 ft.
            469.5 ft.

            469.5 ft. div. by 2.31 = 203.2 psi

---

12. **Answer**   34.3 psi

Draw Diagram.   Change pressure to ft.; add elevation difference; change back to psi. (The length of the line does not matter.)

         30 psi x 2.31 = 69.3 ft.

         69.3 ft. + 10 ft. = 79.3 ft.
         div. by 2.31 = 34.3 psi

---

13. **Answer**   48 psi

Draw Diagram. There must be more pressure below. Change 35 psi to ft.; add elevation; change back to psi.

         35 psi x 2.31 = 80.9 ft.

         80.9 ft. + 30 ft. = 110.9 ft. needed below;
         div. by 2.31 = 48 psi

It will take 48 psi, 30 ft. below the faucet to produce 35 psi at faucet.

14. **Answer**   9590.4 lb. force

Covert psi to psf; solve for force.

> 37 psi x 144 sq. in./sq.ft. = 5328 psf
>
> Force = Pressure x Area
>
> lb. = 5328 psf x 1.8 sq. ft.
>
> lb. = 9590.4

---

15. **Answer**   280,800,000 lb. force

Boat is 25 ft. submerged. Change to psf; obtain bottom area; solve for force.

> 25 ft. x 62.4 lb./cu.ft. = 1560 lb./sq.ft.

> A = l x w   Force = Pressure x Area
>
> A = 300 x 600   lb. = 1560 psf x 180,000 sq.ft.
>
> A = 180,000 sq.ft.   lb. = 280,800,000

---

16. **Answer**   97200 lb. force

Draw Diagram.   The two liquids apply force in opposite directions.   Find the average pressure of each on the wall; subtract one from the other; solve for force.

Oil:
Top pressure = 0

Bottom pressure = .433 lb./sq.in./ft. x .77 x 6 ft. = 2 psi

Average Pressure = 1 psi

Water:
Top pressure = 0

Bottom pressure = $\dfrac{15 \text{ ft.}}{2.31}$ = 6.5 psi

Average pressure = 3.25 psi

3.25 psi (water) - 1 psi (oil) = 2.25 psi pressure from the water side

2.25 psi x 144 sq.in./sq.ft. = 324 psf

Force = Pressure x Area

lb. = 324 psf x 20 ft. x 15 ft.

lb. = 97200

17. **Answer**   93500 lb. force

Draw Diagram. The force is along the entire 200 ft. of length, on the slanted side. Change ft. of water to psf.; solve for force.

Pressure = Weight x Height

psf = 62.4 lb./cu.ft. x 3 ft.

psf = 187.2 at bottom

psf = 0 at top

psf = 93.6 average pressure

Side of triangle can be determined by Pythagorean theorem from geometry $(C^2 = A^2 + B^2)$.

Side = 5 ft. (This is a 3-4-5 right triangle.)

Force = Pressure x Area

lb. = 93.6 x 5 ft. x 200 ft.

lb. = 93500

---

18. **Answer**   19141056 lb. force

Draw Diagram. Set up dimensions; because of 45° slope, water depth is the same as width of water surface in the outer triangles (76 ft.); both triangles can be placed together to form a square 76 ft. x 76 ft.

Solve for average pressure on the dam; solve for force.

Pressure at top = 0

Pressure at bottom = $\dfrac{76 \text{ ft.}}{2.31}$ = 32.9 psi

Average pressure = 16.5 psi x 144 sq.in./sq.ft. = 2376 psf

Solve for area.

Mid section: 76 ft. x 30 ft. = 2280 sq.ft.

Ends together 76 ft. x 76 ft. = 5776 sq.ft.

2280 sq.ft. + 5776 sq.ft. = 8056 sq.ft.

Force = Pressure x Area

lb. = 2376 psf x 8056 sq.ft.

lb. = 19141056 lb.

19.   **Answer**   290,984 lb. force

Draw Diagram.   Find average pressure on heater walls; solve for area of heater walls; solve for force.

Pressure at top of heater = 44 ft. elevation to storage tank plus 15 ft. water depth in it = 59 ft.

$\dfrac{59 \text{ ft.}}{2.31}$ = 25.5 psi

Pressure at bottom of heater = 15 ft. + 50 ft. = $\dfrac{65 \text{ ft.}}{2.31}$ = 28.1 psi

Average pressure = 26.8 psi

Average pressure = 3859.2 psf

Area of heater walls = circumference x height

Area = pi x diam. x height

Area = 3.14 x 4 x 6 ft.

Area = 75.4 sq.ft.

Force = Pressure x Area

lb. = 3859.2 psf x 75.4 sq.ft.

lb. = 290984

---

20.   **Answer**   A. 5.2 psi
                   B. 112320 lb. force

A.   Pressure is equivalent to depth.

$\dfrac{12 \text{ ft.}}{2.31}$ = 5.2 psi

B.   Maximum pressure registers on tank bottom surface as force.

5.2 psi x 144 sq.in./sq.ft. = 748.8 psf

Force = Pressure x Area

lb. = 748.8 psf x 15 ft. x 10 ft.

lb. = 112320

---

21.   **Answer**   A. 10.8 psig
                   B. 25.5 psia
                   C. 1558 psfg
                   D. 3672 psfa

A. $\dfrac{25 \text{ ft.}}{2.31 \text{ ft./psi}} = 10.8$ psig

B. 10.8 psig + 14.7 psi (atmospheric) = 25.5 psia

C. 10.8 psig x 144 sq.in./sq.ft. = 1558 psfg

D. 25.5 psia x 144 sq.in./sq.ft. = 3672 psfa

---

22. **Answer**   5.4 sq.ft.

Calculate thrust: enter pressures as psf.

$$21600 \text{ psf (150 psi) test pressure}$$
$$+\ \underline{14400} \text{ psf (100 psi) water hammer}$$
$$36000 \text{ psf}$$

Thrust = 2T x A x sin ½ Θ

Thrust = 2 x 36000 psf x .785 x 1² x sin 22.5°

Thrust = 21629 lb.

Calculate bearing face area:

$$\text{Area} = \frac{\text{Thrust}}{\text{Bearing Strength (soil)}}$$

$$\text{Area} = \frac{21629 \text{ lb.}}{4000 \text{ lb./sq.ft.}}$$

Area = 5.4 sq.ft.

---

23. **Answer**   A. 3.2 ft.
    B. 368.6 tons additional needed

A. Change tons to pounds; solve for psi pressure of box on water; convert to ft. That's how deep it will sink. When the pressure of the box on the water is equal to the pressure of the water up on the box, it will stop sinking.

80 tons x 2000 lb./ton = 160,000 lb.
This entire weight is resting on its bottom surface.

Bottom surface = 40 ft. x 20 ft. = 800 sq.ft.

That's $\dfrac{160,000 \text{ lb.}}{800 \text{ sq.ft.}} = 200$ lb./sq.ft. = 1.39 psi

1.39 psi x 2.31 = 3.21 ft. (pressure of box on water - in ft. of water)

At this depth underwater, the water pressure pushing upward on the box will be the same. It will sink no further.

B.   To sink it to the bottom, need enough weight to sink it to 18 ft. depth. Make
     a ratio:

$$\frac{80 \text{ tons}}{3.21 \text{ ft. depth}} = \frac{x \text{ tons}}{18 \text{ ft. depth}}$$

x = 448.6 tons total needed to sink it to 18 ft. depth

```
  448.6 tons wt. needed
-  80   tons wt. of box
  368.6 tons additional needed
```

---

24.   **Answer**   8% will show above surface.

An object will rest in water with its top just barely submerged, if its specific gravity
is the same as that of water.

The volume that remains above water in an object of lesser specific gravity, is
equivalent to the ratio of the two specific gravities.

$$\frac{57.2 \text{ lb./cu.ft. (iceberg)}}{62.4 \text{ lb./cu.ft. (water)}} = .92 = 92\% \text{ submerged}$$

8% above water

---

25.   **Answer**   A. Water must be 3.6 ft. deep above ground to float the house.
                   B. Will leak before it floats.

A.   Find pressure of house on its bottom surface; convert to ft. for depth of water
     needed to suspend house.

House bottom surface area = 35 ft. x 60 ft. x 144 sq.in./sq.ft. =
302400 sq.in.

$$\frac{1,250,000 \text{ lb. (wt. of house)}}{302,400 \text{ sq. in}} = 4.13 \text{ psi}$$

This is equivalent to 9.6 ft. of water.

With this depth of groundwater, the house will start to float. But the house is only
6 ft. below grade, so there would have to be a 3.6 ft. deep flood in the yard to float
the house.

B.   To determine leaking point, convert pressure resistance of paint to equivalent
     water pressure in feet.

Resistance of paint = 96 inch column of oil (SG = .8)

96 inches = 8 ft.

.433 lb./ft./sq.in. (wt. of water) x .8 (SG oil) = .346 lb./ft./sq.in. oil

.346 lb./ft./sq.in. oil x 8 ft. column = 2.77 psi oil pressure = paint resistance.

The paint will hold up to 2.77 psi pressure from the groundwater.

The house will float at 4.13 psi pressure from the groundwater.

The house will leak before it floats.

chapter 4

1. **Answer**   .5 ft.

Change gpm to cfs; solve for Velocity; solve for Velocity Head.

2000 gpm x 1440 min./day = 2880000 gpd = 2.88 MGD

2.88 MGD x 1.55 cfs/MGD = 4.46 cfs.

$Q = A \quad V$

$4.46 = .785 \times 1^2 \, V$

$5.67 \text{ ft./sec.} = V$

$Vh = \dfrac{V^2}{2g} = \dfrac{5.67^2}{64.4}$

$Vh = .5$ ft.

2. **Answer**   A. 20 ft.
   B. .19 ft.
   C. 250 ft.
   D. 270.19 ft.

Draw Diagram. Solve for Velocity Head; add heads.

$Ph = 20$ ft = depth of water in the tank.

$Vh = \dfrac{V^2}{2g} = \dfrac{3.5^2}{64.4} = .19$ ft.

$Z = 250$ ft.

Total Head = Hp + Vh + Z

Total Head = 20 + .19 + 250

Total Head = 270.19 ft.

3. **Answer**   A. 192.5 ft.
                   B. 6.2×10^{-8} ft.

A. Change pressure to feet of water; adjust for specific gravity.

> 100 psi x 2.31 ft./psi = 231 ft. (If this liquid were water, this is Ph.)
>
> But SG = 1.2   so $\frac{231}{1.2}$ = 192.5 ft.

B. Change ml./sec. to cfs; solve for velocity; solve for Vh.

> $\frac{5.08 \text{ ml./sec.}}{3785 \text{ ml./gal.}}$ = .0013 gal./sec.
>
> $\frac{.0013 \text{ gal./sec.}}{7.48 \text{ gal./cu.ft.}}$ = .00018 cfs
>
> Q = A   V
>
> .00018 = .785 x $.33^2$ V
>
> .002 ft./sec. = V
>
> Vh = $\frac{V^2}{2g}$ = $\frac{.002^2}{64.4}$ = 6.2×10^{-8}

---

4. **Answer**   .81 ft.

Draw Diagram. Flow is steady, and runs through both
pipes arranged in series. Solve for velocity, then Vh.

> Q = A   V
>
> 1.42 = .785 x $.5^2$ V
>
> 7.24 ft./sec. = V
>
> Vh = $\frac{V^2}{2g}$ = $\frac{7.24^2}{64.4}$ = .81 ft.

---

5. **Answer**   42.9 psi

Draw Diagram. Solve for flow; solve for Vel. in 6 inch
pipe; solve for Vh in both; change psi to feet of water;
using Bernoulli's formula, solve for Ph in 6 inch pipe.

> Q = A   V
>
> Q = .785 x 1.167^2 x 20
>
> Q = 21.4 cfs

$$Q = A \quad V$$

$$21.4 = .785 \times .5^2 \, V$$

$$108.9 \text{ ft./sec.} = V$$

$$Vh = \frac{V^2}{2g}$$

$$Vh = \frac{20^2}{64.4}$$

$$Vh = 6.2 \text{ ft.}$$

$$Vh = \frac{V^2}{2g}$$

$$Vh = \frac{108.9^2}{64.4}$$

$$Vh = 184.3 \text{ ft.}$$

$$Ph_1 + Vh_1 + Z_1 = Ph_2 + Vh_2 + Z_2$$

$$277.2 + 6.2 + 0 = Ph + 184.3 + 0$$

$$99.13 \text{ ft.} = Ph$$

$$42.9 \text{ psi} = \text{pressure}$$

6. **Answer**   1.4 cfs

Draw Diagram. Change pressure to feet of water; using Bernoulli's, solve for Vh, then velocity, then flow.

$$P = W \times H$$

$$312 \text{ lb./sq.ft.} = 62.4 \text{ lb./cu.ft.} \times \text{ft.}$$

$$5 \text{ ft.} = H$$

$$Th = Ph + Vh + Z$$

$$19.3 \text{ ft.} = 5 \text{ ft.} + Vh + 13.5 \text{ ft.}$$

$$.8 \text{ ft.} = Vh$$

$$Vh = \frac{V^2}{2g}$$

$$.8 = \frac{V^2}{64.4}$$

$$7.2 \text{ ft./sec.} = V$$

$$Q = A \quad V$$

$$Q = .785 \times .5^2 \times 7.2$$

$$Q = 1.4 \text{ cfs}$$

7. **Answer** 7 cfs

Draw Diagram. Change pressure to feet of water; setting up Bernoulli's formula it appears that not enough information is given; both Vh's are unknown; however, a relationship can be established between the velocities of the two pipes; using different assumed flows to solve for velocity, it is found that the velocity in the smaller pipe will always be four times the velocity in the larger pipe. Obtain velocity; then solve for flow.

50 psi x 2.31 ft./psi = 115.5 ft.

10 psi x 2.31 ft./psi = 23.1 ft.

$$Ph_1 + Vh_1 + Z_1 = Ph_2 + Vh_2 + Z_2$$

$$115.5 + \frac{V^2}{64.4} + 0 = 23.1 + \frac{(4V)^2}{64.4} + 0$$

20 ft./sec. = V

79.6 ft./sec. = 4V

Q = A   V

Q = .785 x .67² x 20

Q = 7 cfs

---

8. **Answer** 1 cfs

Solve for velocity; solve for flow.

$$Vh = \frac{V^2}{2g}$$

$$.44 = \frac{V^2}{64.4}$$

5.3 ft./sec. = V

Q = A   V

Q = .785 x .5² x 5.3

Q = 1 cfs

---

9. **Answer** 17.6 ft./sec.

Draw Diagram. Solve for Vh, then velocity.

$$Ph_1 + Vh_1 = Ph_2 + Vh_2$$

$$46.2 + \frac{30^2}{64.4} = 55.4 + Vh_2$$

$$4.8 \text{ ft.} = Vh_2$$

$$Vh = \frac{V^2}{2g}$$

$$4.8 = \frac{V^2}{64.4}$$

$$17.6 \text{ ft./sec.} = V$$

---

10. **Answer**   A. 7.1 ft./sec.
         B. 40.5 ft./sec.
         C. 10.6 psi

Draw Diagram. Head loss is negligible because
of short distance.

A. Solve for velocity.

$$Q = A \quad V$$

$$1.4 \text{ cfs} = .785 \times .5^2 \, V$$

$$7.1 \text{ ft./sec.} = V$$

B. Solve for velocity in nozzle.

$$Q = A \quad V$$

$$1.4 = .785 \times .21^2 \, V$$

$$40.5 \text{ ft./sec.} = V$$

C. Use Bernoulli's formula to solve for Ph in pipe.

Nozzle:
$$Vh = \frac{V^2}{2g}$$

$$Vh = \frac{40.5^2}{64.4} = 25.4 \text{ ft.}$$

Pipe:
$$Vh = \frac{V^2}{2g}$$

$$Vh = \frac{7.1^2}{64.4} = .8 \text{ ft.}$$

$$Ph_1 + Vh_1 + Z_1 = Ph_2 + Vh_2 + Z_2$$

$$Ph_1 + .8 + 0 = 0 + 25.4 + 0$$

$$Ph = 24.6 \text{ ft.}$$

$$\text{Pressure} = 10.6 \text{ psi}$$

11. **Answer**   60 psi

Draw Diagram.   Change pressure to feet of water; use
Bernoulli's formula to solve for exit pressure head; change
to psi. (Vh at both points is the same.)

$$Ph_1 + Z_1 = Ph_2 + Z_2 + HL$$

$$150.2 + 0 = Ph_2 + 4 + 8$$

$$138.2 \text{ ft.} = Ph_2$$

60 psi = Pressure at effluent

This problem is easy enough to solve from visual inspection once the diagram is
drawn, but use of Bernoulli's formula eliminates chance of error.

12. **Answer**   9.2 ft.

Draw Diagram. Use Bernoulli's formula. Recognize that
though all Ph is lost, some has been converted to velocity
head, and some has been lost due to pipe friction. We are
solving for that friction.

Solve for velocity; then Vh.

$$Q = A \quad V$$

$$25 = .785 \times 1^2 \, V$$

$$31.9 = V$$

$$Vh = \frac{V^2}{2g}$$

$$Vh = \frac{31.9^2}{64.4} = 15.8 \text{ ft.}$$

$$Ph_1 + Vh_1 + Z_1 = Ph_2 + Vh_2 + Z_2 + HL$$

$$25 + 0 + 0 = 0 + 15.8 + 0 + HL$$

$$9.2 \text{ ft.} = HL$$

13. **Answer**   0 psi

Pressure discharged is flowing free, and has only velocity head.

14. **Answer**   8 ft./sec.

Draw Diagram. Change pressure to feet of water; use Bernoulli's formula to solve for Vh, then velocity.

$$13 \text{ psi} \times 2.31 \text{ ft./psi} = 30 \text{ ft.}$$

$$TH = Ph + Vh + Z$$

$$131 = 30 + Vh + 100$$

$$1 \text{ ft.} = Vh$$

$$Vh = \frac{V^2}{2g}$$

$$1 = \frac{V^2}{64.4}$$

$$8 \text{ ft./sec.} = V$$

15. **Answer**   71.4 ft.

Draw Diagram. Change pressure to ft. of water; by visual inspection we can see that all the pressure energy is lost as the water exits the pipe. Vh remains the same at exit, and cancels out of Bernoulli's formula, if used.

$$33.5 \text{ psi} \times 2.31 \text{ ft./psi} = 77.4 \text{ ft.}$$

$$Ph_1 + Z_1 = Ph_2 + Z_2 + HL$$

$$77.4 + 0 = 0 + 6 + HL$$

$$71.4 \text{ ft.} = HL$$

16. **Answer**   14 ft.

Draw Diagram. The reading on the discharge gage registers the pressure in the discharge pipe; in feet of water, this is the lift. Change pressure to feet.

$$45 \text{ psi} \times 2.31 \text{ ft./psi} = 104 \text{ ft.}$$

But the water is only being lifted 90 ft. The difference is the pressure lost inpipe because of friction.

$$104 \text{ ft.} - 90 \text{ ft.} = 14 \text{ ft. Head Loss}$$

17. **Answer**   A. 80 ft.
                 B. 187898 gal.

Draw Diagram.

A. Water level in #1 is 120 ft. above pump. Pump must
    provide pressure from this level up to top of #2.

                 200 ft. - 120 ft. = 80 ft.

B. Reservoir #2 (diam. 40 ft., depth 20 ft.) same size as #1. Solve for volume.
    Change to gallons.

                 Vol. = .785 x d^2 x h
                 Vol. = .785 x 40^2 x 20
                 Vol. = 25120 cu.ft.
                 Vol. = 187898 gallons

---

18. **Answer**   119.3 ft.

Draw Diagram.   Change head loss to feet of water;
change house pressure to feet of water; use Bernoulli's
formula to solve for tank depth (Ph_1). Velocity Heads
cancel.

                 43.3 psi x 2.31 ft./psi = 100 ft.
                 30 psi x 2.31 ft./psi = 69.3 ft.

                 Ph_1 + Vh_1 + Z_1 = Ph_2 + Vh_2 + Z_2 + HL
                 Ph_1 + 50 = 69.3 + 0 + 100
                 Ph_1 = 119.3 ft.

---

19. **Answer**   35.3 ft.

Draw Diagram.   Change pressures to feet; solve for
velocity at point 2; solve for Vh's at both points; use
Bernoulli's formula to solve for Head Loss.

                 60 psi x 2.31 ft./psi = 138.6 ft.
                 40 psi x 2.31 ft./psi = 92.4 ft.

                 A   V = A   V
                 .785 x 6^2 x 2 = .785 x 3^2 V
                 8 ft./sec. = V

$$Vh = \frac{V^2}{2g}$$

$$Vh = \frac{8^2}{64.4} = 1 \text{ ft.}$$

$$Vh = \frac{V^2}{2g}$$

$$Vh = \frac{3^2}{64.4} = .14 \text{ ft.}$$

$$Ph_1 + Vh_1 + Z_1 = Ph_2 + Vh_2 + Z_2 + HL$$

$$138.6 + .14 + 0 = 92.4 + 1 + 10 + HL$$

$$35.3 \text{ ft.} = HL$$

20. **Answer** 51 psi

Draw Diagram. Change pressures to feet of water; solve for velocities at both points, then velocity head at point 2; use Bernoulli's formula to solve for pressure at point 2.

48 psi x 2.31 ft./psi = 111 ft.

$$\frac{4.3 \text{ inches}}{12 \text{ inches/ft.}} = .36 \text{ ft. HL}$$

$$Vh = \frac{V^2}{2g}$$

$$.63 = \frac{V^2}{64.4}$$

6.4 ft./sec. V

A V = A V

.785 x .67² x 6.4 = .785 x .83² V

4.2 ft./sec. = V

$$Vh = \frac{V^2}{2g}$$

$$Vh = \frac{4.2^2}{64.4} = .27 \text{ ft.}$$

$$Ph_1 + Vh_1 + Z_1 = Ph_2 + Vh_2 + Z_2 + HL$$

$$111 + .63 + 7 = Ph_2 + .27 + 0 + .36$$

$$118 \text{ ft.} = Ph_2$$

51 psi = Pressure

21. **Answer**   2.09 cfs

Pressure differential on pitot gage is the Vh. Change it to feet of water. Then to solve for flow we must solve for velocity, then flow. HL restricts vel. (or flow) by 10%. Reduce flow amount by 10%.

$$4 \text{ psi} \times 2.31 \text{ ft./psi} = 9.24 \text{ ft.}$$

$$Vh = \frac{V^2}{2g}$$

$$9.24 = \frac{V^2}{64.4}$$

$$24.4 \text{ ft./sec.} = V$$

$$Q = A \ V$$

$$Q = .785 \times .33^2 \times 24.4$$

$$Q = 2.09 \text{ cfs}$$

---

22. **Answer**   A. 1.57 cfs
                  B. 161.2 ft.

A. Solve for flow.

$$Q = A \ V$$

$$Q = .785 \times .5^2 \times 8$$

$$Q = 1.57 \text{ cfs}$$

B. Draw Diagram. Change pressures to feet of water. Solve for velocity heads on both suction and discharge sides. Solve for Total Head on each side. Subtract suction head from discharge head to yield head which pump is providing.

<u>Suction Side</u>
$$Q = A \ V$$

$$1.57 = .785 \times .67^2 \times V$$

$$4.46 \text{ ft./sec.} = V$$

$$Vh = \frac{V^2}{2g}$$

$$Vh = \frac{4.46^2}{64.4} \text{ ft.}$$

$$Vh = .3 \text{ feet}$$

<u>Discharge Side</u>
$$Vh = \frac{V^2}{2g}$$

$$Vh = \frac{8^2}{64.4}$$

$$Vh = 1 \text{ ft.}$$

<u>Suction Side</u>
$$TH = Ph + Vh + Z$$

$$TH = (-13.9) + .3 + 0$$

$$TH = -13.6 \text{ ft.}$$

<u>Discharge Side</u> - **(at gage)**
$$TH = Ph + Vh + Z$$

$$TH = 138.6 + 1 + 8$$

$$TH = 147.6 \text{ ft.}$$

Discharge Head - Suction Head = Pumping Head

$$147.6 - (-13.6) = 161.2$$

Pump provides 161.2 ft. of head

---

23. **Answer**   8 ft.

Draw Diagram. The jet will travel up a distance that is equivalent to the velocity head; this is the only energy it has at this point. Change gpm to cfs; solve for velocity at exit; solve for VH.

2000 gpm x 1440 min./day = 2880000 gpd = 2.88 MGD

2.88 MGD x 1.55 cfs/MGD = 4.46 cfs

$$Q = A \ V$$

$$4.46 = .785 \times .5^2 \ V$$

$$22.7 \text{ ft./sec.} = V$$

$$Vh = \frac{V^2}{2g}$$

$$Vh = \frac{22.7^2}{64.4} = 8 \text{ ft.}$$

Bernoulli's formula can be used to prove that the elevation distance of travel is equal to Vh.

$$Ph_1 + Vh_1 + Z_1 = Ph_2 + Vh_2 + Z_2 + HL$$

$$0 + 8 + 0 = 0 + 0 + Z + 0$$

$$8 \text{ ft.} = Z_2$$

24. **Answer** 73 psi

Draw Diagram. Set point 1 at pump discharge, and point 2 at end of nozzle.

Use Bernoulli's formula to determine pressure at pump. Solve for Vh at pump first, then insert all quantities into Bernoulli's formula.

$$Vh = \frac{V^2}{2g}$$

$$80 = \frac{V^2}{64.4}$$

$$V = 71.8 \text{ ft./sec}$$

$$A \ V = A \ V$$

$$.785 \times .5^2 \ V = .785 \times .083^2 \times 71.8$$

$$V = 1.98 \text{ ft./sec}$$

$$Vh = \frac{V^2}{2g} = \frac{1.98^2}{64.4}$$

$$Vh = .061 \text{ ft.}$$

Velocity head at this point could be eliminated; it is negligible.

$$Ph_1 + Vh_1 + Z_1 = Ph_2 + Vh_2 + Z_2 + HL$$

$$Ph_1 + .061 + 0 = 0 + 80 + 60 + 28$$

$$Ph_1 = 168 \text{ ft.}$$

$$\text{Pressure} = 73 \text{ psi}$$

Vh at point 2 is 80 ft. (the pump is throwing water 80 ft. up)

---

25. **Answer** A. 6 inch diameter
    B. -6.5 ft. pressure at summit

Draw Diagram.

Solve for Ph at summit (point 1 at water surface, point 2 at summit of siphon); solve for HL in downward length of siphon; this equal 1Vh; solve for velocity; then diameter.

$$Ph_1 + Z_1 = Ph_2 + Z_2 + HL$$

$$0 + 214 = Ph_2 + 219 + 1.5$$

$$Ph_2 = -6.5 \text{ ft.}$$

$$Ph_2 + Z_2 = Ph_3 + Z_3 + HL$$

$$-6.5 + 219 = 0 + 0 + HL$$

$$212.5 \text{ ft.} = HL$$

$$212.5 \text{ ft.} = 1Vh$$

$$Vh = \frac{V^2}{2g}$$

$$212.5 = \frac{V^2}{64.4}$$

$$117 \text{ ft./sec.} = V$$

$$Q = A \ V$$

$$20 = .785d^2 \times 117$$

$$.47 \text{ ft.} = d$$

$$6 \text{ inches} = d$$

chapter 5

1. **Answer**   130 ft.

Draw Diagram.   Pump lifts water to storage tank.   Discharge gage registers Dynamic Discharge Head (lift plus losses).   Change pressure to feet of water; solve for head loss.

$$95 \text{ psi} \times 2.31 \text{ ft./psi} = 220 \text{ ft.}$$

$$DDH = Lift + Losses$$

$$220 = 90 + x$$

$$130 \text{ ft.} = x$$

2. **Answer**   10 ft.

Draw Diagram.   Total Dynamic Head includes all the work the pump is doing (lift & losses on both suction and discharge sides).   Dynamic Discharge Head = lift & losses on the discharge side.   Solve for suction lift.

$$TDH = DSL + DDH$$

$$150 = (x + 10) + 130$$

$$10 \text{ ft.} = x$$

3. **Answer** 18.6 hp

$$\frac{\text{gpm x hd}}{3960} = \text{whp}$$

$$\frac{1840 \times 40}{3960} = \text{whp}$$

$$18.6 = \text{whp}$$

4. **Answer** 17.2 psi

Solve for head; change to pressure.

$$\frac{\text{gpm x hd}}{3960} = \text{whp}$$

$$\frac{1200 \times \text{hd}}{3960} = 12$$

$$\text{hd} = 40 \text{ ft.}$$

$$\frac{40 \text{ ft.}}{2.31 \text{ ft./psi}} = 17.2 \text{ psi}$$

5. **Answer** 46 ft.

$$\frac{\text{gpm x hd}}{3960} = \text{whp}$$

$$\frac{430 \times \text{hd}}{3960} = 5$$

$$\text{hd} = 46 \text{ ft.}$$

6. **Answer** A. 19.8 hp
            B. 26.4 hp
            C. 32.2 hp

A. $\dfrac{\text{gpm x hd}}{3960} = \text{whp}$

$$\frac{825 \times 95}{3960} = \text{whp}$$

$$19.8 = \text{whp}$$

B. $\dfrac{\text{whp}}{\text{pump eff.}} = \text{bhp}$

$$\frac{19.8}{.75} = \text{bhp}$$

$$26.4 = \text{bhp}$$

C. $\dfrac{bph}{\text{motor eff.}}$ = mhp

$\dfrac{26.4}{.82}$ = mhp

32.2 = mhp

---

7. **Answer**   6.2 hp

Change pressure to feet of water; solve for motor horsepower.

45 psi x 2.31 ft./psi = 104 ft.

$\dfrac{\text{gpm x ft.}}{3960}$ = whp

$\dfrac{200 \text{ x } 104}{3960}$ = whp

5.3 = whp

$\dfrac{whp}{\text{pump eff. x motor eff.}}$ = mhp

$\dfrac{5.3}{.94 \text{ x } .90}$ = mhp

6.2 = mhp

---

8. **Answer**   49 hr./week

Change watts to Kilowatts; solve for hours.

$\dfrac{2238 \text{ watts}}{1000 \text{ watts/kw}}$ = 2.238 Kw

Kw x hr. x cost/Kw.hr. = Total cost

2.238 x hr. x .015 = 1.65

hr./week = 49

---

9. **Answer**   1257 gpm

"Operating against a pressure of" means "Total Dynamic Head" - the work the pump is doing. A horsepower rating given as the property of the pump is the Brake Horsepower. Solve for whp; solve for gpm.

$\dfrac{whp}{\text{pump eff.}}$ = bhp

$\dfrac{whp}{.88}$ = 25

whp = 22

$$\frac{\text{gpm x hd}}{3960} = \text{whp}$$

$$\frac{\text{gpm x 69.3}}{3960} = 22$$

$$\text{gpm} = 1257$$

10. **Answer**   3.6 ft.

Solve for whp (it is the power needed to move the water); solve for TDH (work the pump has to do-includes suction and discharge); solve for suction lift.

$$\frac{\text{whp}}{\text{pump eff.}} = \text{bhp}$$

$$\frac{\text{whp}}{.86} = 6.5$$

$$\text{whp} = 5.6$$

$$\frac{\text{gpm x hd}}{3960} = \text{whp}$$

$$\frac{590 \text{ x hd}}{3960} = 5.6$$

$$\text{hd} = 37.6 \text{ ft. (TDH)}$$

Suction Lift + Other Heads = TDH

Suction Lift + 34 = 37.6

Suction Lift = 3.6 ft.

11. **Answer**   81% efficient

Solve for WHP; solve for pump efficiency.

$$\frac{\text{gpm x hd}}{3960} = \text{whp}$$

$$\frac{355 \text{ x } 41}{3960} = \text{whp}$$

$$3.7 = \text{whp}$$

$$\frac{\text{whp}}{\text{motor eff. x pump eff.}} = \text{mhp}$$

$$\frac{3.7}{.91 \text{ x pump eff.}} = 5$$

$$\text{pump eff.} = .81$$

12. **Answer** $14.59

$$\frac{\text{gpm x hd}}{3960} = \text{whp}$$

$$\frac{700 \text{ x } 82}{3960} = \text{whp}$$

$$14.5 = \text{whp}$$

$$\frac{\text{whp}}{\text{w to w eff.}} = \text{mhp}$$

$$\frac{14.5}{.83} = \text{mhp}$$

$$17.5 = \text{mhp}$$

17.5 hp x .746 Kw/hp x 14 hr. x $.08/Kw.hr. = $14.59

---

13. **Answer**   67% efficiency

Draw Diagram. Convert Kw to mhp; solve for TDH, then whp; solve for w-w eff.

$$5.5 \text{ Kw/30 min.} = 11 \text{ Kw./hr.}$$

$$\frac{11 \text{ Kw/hr}}{.746 \text{ Kw/hp}} = 14.8 \text{ mhp}$$

$$\frac{\text{gpm x hd}}{3960} = \text{whp}$$

$$\frac{300 \text{ x } 130.4}{3960} = \text{whp}$$

$$9.9 = \text{whp}$$

$$\frac{\text{wph}}{\text{w-w eff.}} = \text{mhp}$$

$$\frac{9.9}{\text{w-w eff.}} = 14.8$$

$$\text{w-w eff.} = .67$$

---

14. **Answer**   36.6 Kw

$$\frac{\text{gpm x hd}}{3960} = \frac{\text{whp}}{\text{w/w eff}} = \text{mhp x .746} = \text{Kw}$$

$$\frac{350 \text{ x } 400}{3960} = \frac{35.35}{.72} = 49 \text{ x .746} = 36.6 \text{ Kw}$$

15. **Answer**   A.   30 ft.
                      B.   16.7

    A.   Drawdown = Static Water Level - Pumping Water Level

        Drawdown = (52 psi x 2.31 ft./psi) - (39 psi x 2.31 ft./psi)

        Drawdown = 30 ft.

    B.   Sp.Cap. = gpm/ft. drawdown

        Sp.Cap. = 500/30

        Sp.Cap. = 16.7

---

16. **Answer**   A.   100 ft.
                      B.   1.5

Note that this data records the Static Water Level as the distance from ground surface to the water table, not the depth of static water in the well.

    A.   Drawdown = Pumping Water Level - Static Water Level

        Drawdown = 500 - 400

        Drawdown = 100 ft.

    B.   Sp.Cap. = gpm/ft. drawdown

        Sp.Cap. = 150/100

        Sp.Cap. = 1.5

---

17. **Answer**   52 psi

Draw Diagram. Pumping rate and water table of 200 ft. below surface not needed for this solution. Change pressure to feet of water; add to Drawdown; change to psi.

           29.5 psi x 2.31 ft./psi = 68.2 ft.

           68.2 ft. + 52 ft. = 120.2 ft.
                   (water depth-pump off)

$$\frac{120.2 \text{ ft.}}{2.31 \text{ ft./psi}} = 52 \text{ psi}$$

18. **Answer**   A. 73 ft.
                     B. 53 ft.

A. Gage reading (pump off) is equivalent to depth of standing water in the well. Subtract from well depth.

$$55 \text{ psi} \times 2.31 \text{ ft./psi} = 127 \text{ ft. (SWL)}$$

$$200 \text{ ft.} - 127 \text{ ft.} = 73 \text{ ft. to water}$$

B. Gage reading (pump on) is equivalent to pumping water level. Subtract from Static Water Level.

$$32 \text{ psi} \times 2.31 \text{ ft./psi} = 74 \text{ ft.}$$

$$127 \text{ ft.} - 74 \text{ ft.} = 53 \text{ ft. Drawdown}$$

19. **Answer**   235 min.

Draw Diagram. Solve for pumping rate; solve for volume of empty half of tank; divide by pumping rate.

$$100 \text{ gpm in} - 50 \text{ gpm out} = 50 \text{ gpm pumping rate}$$

$$\text{Vol.} = .785 \; d^2 \; h$$

$$\text{Vol.} = .785 \times 20^2 \times 5$$

$$\text{Vol.} = 1570 \text{ cu.ft.}$$

$$\text{Vol.} = 11744 \text{ gallons}$$

$$\frac{11744 \text{ gal.}}{50 \text{ gal./min.}} = 235 \text{ minutes to fill}$$

20. **Answer**   718 gpm

Draw Diagram. Solve for cfs in; solve for cfs drop; rise + drop = gpm pumped out.

Rise = 2 ft./min. 46 seconds; find volume of that 2 ft. depth

$$\text{Vol.} = .785 \; d^2 \; h$$

$$\text{Vol.} = .785 \times 10^2 \times 2$$

$$\text{Vol.} = 157 \text{ cu.ft.}$$

$$\frac{157 \text{ cu.ft.}}{1.77 \text{ min.}} = 88.7 \text{ cu.ft./min.} = 1.5 \text{ cfs } \text{in}$$

Drop = 4 inches/4 min.; find that volume; change to cfs drop.

$$\text{Vol.} = .785 \ d^2 \ h$$

$$\text{Vol.} = .785 \times 10^2 \times .33$$

$$\text{Vol.} = 26 \ \text{cu.ft.}$$

$$\frac{26 \ \text{cu.ft.}}{4 \ \text{min.}} = 6.48 \ \text{cu.ft./min.} = .1 \ \text{cfs}$$

$$\text{Rise} + \text{Drop} = \text{Pumping rate}$$

$$1.5 + .1 = 1.6 \ \text{cfs pumping rate}$$

$$1.6 \ \text{cfs} \times 7.48 \ \text{gal./cu.ft.} \times 60 \ \text{sec./min.} = 718 \ \text{gpm}$$

---

21. **Answer**   122.7 gpm

Not all information needed.

$$\frac{3680 \ \text{gal.}}{30 \ \text{min.}} = 122.7 \ \text{gpm}$$

---

22. **Answer**   7.3 psi

Solve for depth of water in tank after pumping; change to pressure.

$$200 \ \text{gpm} \times 60 \ \text{min./hr.} \times 4 \ \text{hr.} = 48{,}000 \ \text{gal.} = 6417 \ \text{cu.ft.}$$

$$\text{Vol.} = .785 \ d^2 \times h$$

$$6417 = .785 \times 22^2 \times h$$

$$17 \ \text{ft.} = h$$

$$\frac{17 \ \text{ft. depth}}{2.31 \ \text{ft./psi}} = 7.3 \ \text{psi}$$

---

23. **Answer**   2.8 hours

$$500 \ \text{gpm} \ \underline{in} - 200 \ \text{gpm} \ \underline{out} = 300 \ \text{gpm filling rate}$$

empty 2/3 of tank = 50,000 gal.

$$\frac{50{,}000 \ \text{gal.}}{300 \ \text{gpm filling}} = 166.7 \ \text{minutes to fill tank (2.8 hours)}$$

---

24. **Answer**   Pump will run dry. More water is being pumped out than is coming in.

It is almost impossible to determine a constant regular influent rate for a wastewater lift station. Best solution is to set up pump to respond to water level control devices.

.1 cfs <u>in</u> x 7.48 gal./cu.ft. x 60 sec./min. x 60 min./hr. = 2693 gal./hr. <u>in</u>.

150 gpm <u>out</u> x 20 min/hr. = 3000 gal./hr. <u>out</u>.

25. **Answer**   90% of design efficiency

Solve for gpm pumping rate.

Inflow + Drop = Pumping Rate

Influent 110 cfm = 1.83 cfs

Drop 15 inches/4 min. 20 sec. Find volume of drop; change to cfs drop rate.

Vol. = l x w x h

Vol. = 11 x 11 x 1.25

Vol. = 151.25 cu.ft. drop - in 4 min. 20 sec.

$$\frac{151.25 \text{ cu.ft.}}{4.33 \text{ min.}} = 24.9 \text{ cu.ft./min.} = .58 \text{ cfs}$$

1.83 cfs <u>in</u>
+ .58 cfs out
  2.41 cfs pumping out

2.41 cfs x 7.48 gal./cu.ft. x 60 sec./min. = 1084 gpm actually being pumped

$$\frac{1084 \text{ gpm pumping rate}}{1200 \text{ gpm design rate}} = .9 = 90\%$$

chapter 6

1. **Answer**   .00001 ft./ft.

Change gpm to cfs; solve for slope.

$$\frac{800 \text{ gpm}}{60 \text{ sec./min x } 7.48 \text{ gal./cu.ft.}} = 1.78 \text{ cfs}$$

$Q = .435 \ C \ d^{2.63} \ s^{.54}$

$1.78 = .435 \ \ 100 \ \ 3^{2.63} \ s^{.54}$

$.00001 = s$

2. **Answer**   1.2 ft.

Rough concrete $C=110$; 30 inches = 2.5 ft.

$Q = .435 \ C \ d^{2.63} \ s^{.54}$

$10 = .435 \ \ 110 \ \ 2.5^{2.63} \ s^{.54}$

$.0006 = s$

$s = \dfrac{HL}{L}$

$.0006 = \dfrac{HL}{2000}$

$1.2 \text{ ft.} = HL$

3. **Answer**   .64 ft.

Change MGD to cfs; 10 in. = .83 ft.; Unlined steel $C=140$.

$.8 \text{ MGD x } 1.55 = 1.24 \text{ cfs}$

$Q = .435 \ C \ d^{2.63} \ s^{.54}$

$1.24 = .435 \ \ 140 \ \ .83^{2.63} \ s^{.54}$

$.002 = s$

$HL = .002 \text{ ft./ft. x } 350 \text{ ft.} = .64 \text{ ft.}$

4. **Answer**   37.5 ft.

CIP 40 yr. old C=50

$$Q = .435 \ C \ d^{2.63} \ s^{.54}$$

$$.2 = .435 \ 50 \ .33^{2.63} \ s^{.54}$$

$$.0375 = s$$

$$HL = .0375 \ ft./ft. \times 1000 \ ft. = 37.5 \ ft.$$

5. **Answer**   45.9 ft.

Draw Diagram.   Pressure at house (40 psi) is equivalent to 92.4 ft. With Bernoulli's formula, solve for depth needed; solve for HL; add to depth.

$$Ph_1 + Z_1 = Ph_2 + Z_2 + HL$$

$$Ph + 50 = 92.4 + 0 + HL$$

$$Ph = 42.4 + HL$$

$$Q = .435 \ C \ d^{2.63} \ s^{.54}$$

$$2 = .435 \ 110 \ .83^{2.63} \ s^{.54}$$

$$.007 = s$$

$$HL = .007 \ ft./ft. \times 500 \ ft. = 3.5 \ ft.$$

$$\begin{array}{r} 42.4 \ ft. \\ + \ 3.5 \ ft. \\ \hline 45.9 \ ft. \end{array}$$

6. **Answer**   114.6 psi

Draw Diagram.   10 yr. old AC pipe C=130. Treatment plant must deliver 60 psi plus 100 ft. elevation, plus make up HL; solve for HL; add:

$$4.5 \ MGD \times 1.55 \ cfs/MGD = 6.98 \ cfs$$

$$60 \ psi \times 2.31 \ ft./psi = 138.6 \ ft.$$

$$Ph_1 + Z_1 = Ph_2 + Z_2 + HL$$

$$Ph_1 + 0 = 138.6 + 100 + HL$$

$$Ph_1 = 238.6 + HL$$

$$Q = .435 \ C \ d^{2.63} \ s^{.54}$$

$$6.98 = .435 \ \ 130 \ \ 1.5^{2.63} \ s^{.54}$$

$$.003 = s$$

HL = .003 ft./ft. x 9000 ft. = 26 ft.

238.6 ft. + 26 ft. = 264.6 ft.

$$\frac{264.6 \ ft.}{2.31 \ ft./psi} = 114.6 \ psi$$

7. **Answer**   19.5 psi

Galv. pipe C=120. Solve for flow; solve for HL.

$$Q = A \ V$$

$$Q = .785 \ x \ .5^2 \ x \ 8$$

$$Q = 1.57 \ cfs$$

$$Q = .435 \ C \ d^{2.63} \ s^{.54}$$

$$1.57 = .435 \ \ 120 \ \ .5^{2.63} \ s^{.54}$$

$$.045 = s$$

HL = .045 ft./ft. x 1000 ft. = 45 ft.

$$\frac{45 \ ft.}{2.31 \ ft./psi} = 19.5 \ psi$$

8. **Answer**   18 inch diameter pipe

slope 25 ft./5280 ft. = .0047

lined steel pipe C=150

$$Q = .435 \ C \ d^{2.63} \ s^{.54}$$

$$10 = .435 \ \ 150 \ d^{2.63} \ .0047^{.54}$$

$$1.47 \ ft. = d$$

use 18 inch pipe

9. **Answer**   C = 80

slope = 4 ft./8000 ft. = .0005

20 inches = 1.67 ft.

$$Q = .435 \ C \ d^{2.63} \ s^{.54}$$

$$2.2 = .435 \ C \ 1.67^{2.63} \ .0005^{.54}$$

$$80 = C$$

---

10. **Answer**   A. 73.5 psi
                      B. Velocity is the same at both points.
                          (Flow is steady - area does not change; Q=AV.)

Draw Diagram. Change psi to ft.; change gpm to cfs; solve for HL; subtract.

80 psi x 2.31 ft./psi = 184.8 ft.

$$\frac{1500 \ gpm}{60 \ sec./min. \ x \ 7.48 \ gal./cuft.} = 3.3 \ cfs$$

$$Q = .435 \ C \ d^{2.63} \ s^{.54}$$

$$3.3 = .435 \ 120 \ .83^{2.63} \ s^{.54}$$

$$.015 = s$$

HL = .015 ft./ft. x 1000 ft. = 15 ft.

184.8 ft. (point 1) - 15 ft. (HL) = 169.8 ft. = 73.5 psi

---

11. **Answer**   30.3 psi difference

Draw Diagram. As pipe slopes up,pressure drops-equivalent to 50 ft.; pressure also drops from HL; solve for Q; solve for HL; add to elevation.

Galvanized C=120

$$Q = A \ V$$

$$. Q = .785 \ x \ 1^2 \ x \ 9$$

$$Q = 7.1 \ cfs$$

$$Q = .435 \ C \ d^{2.63} \ s^{.54}$$

$$7.1 = .435 \ 120 \ 1 \ s^{.54}$$

$$.025 = s$$

HL = .025 ft./ft. x 800 ft. = 20 ft.

50 ft. (elev.) + 20 ft. (HL) = 70 ft. pressure difference

70 ft. = 30.3 psi

12. **Answer** .003 ft./ft.

CIP C=130
Find flow in 20 in. pipe; using that flow, solve for s in 16 in. pipe

HGL drop 1 ft./1000 ft. = .001 = s

$Q = .435 \ C \ d^{2.63} \ s^{.54}$

$Q = .435 \ 130 \ 1.67^{2.63} \ .001^{.54}$

$Q = 5.2 \ cfs$

$Q = .435 \ C \ d^{2.63} \ s^{.54}$

$5.2 = .435 \ 130 \ 1.33^{2.63} \ s^{.54}$

$.003 = s$

Note: Since the flow in both pipes is the same, <u>the Total HL will be the same in both</u>. The smaller pipe has a greater HL per foot of pipe - and must then be a shorter pipe.

---

13. **Answer** 3.33 cfs

New concrete C=140

HL = 2 psi/1000 ft. = 4.62 ft./1000 ft. = .0046 ft./ft.

$Q = .435 \ C \ d^{2.63} \ s^{.54}$

$Q = .435 \ 140 \ 1 \ .0046^{.54}$

$Q = 3.33 \ cfs$

---

14. **Answer** .48 cfs

CIP C=130; solve for HL in the new pipe; with narrowed pipe, at the same HL, solve for flow. This is not a steady state system. The flow will change-because the pipe narrowed.

New Pipe
$Q = .435 \ C \ d^{2.63} \ s^{.54}$

$1 = .435 \ 130 \ .33^{2.63} \ s^{.54}$

$.1258 = s$

Narrowed Pipe
$Q = .435 \ C \ d^{2.63} \ s^{.54}$

$Q = .435 \ 130 \ .25^{2.63} \ .1258^{.54}$

$Q = .48 \ cfs$

15. **Answer**   20%

Test it out with 2 pipes; keep all quantities the same - change C by 20%; solve for flow.

---

16. **Answer**   A. 6 inch diam.
                  B. ?

Masonry conduit C=130; change gph to cfs.

$$3900 \text{ gph}/7.48 \times 60 \times 60 = .145 \text{ cfs}$$

A. Solve for diam. First find s. $1.18/570 = .0021$ ft./ft.

$$Q = .435 \ C \ d^{2.63} \ s^{.54}$$

$$.145 = .435 \ 130 \ d^{2.63} \ .0021^{.54}$$

$$.367 \text{ ft.} = d$$

$$4.4 \text{ in.} = d$$

6 in. pipe would be chosen

B. Solve for velocity with 6 inch pipe at this flow.

$$Q = A \ V$$

$$.145 = .785 \times .5^2 \ V$$

$$.739 \text{ ft./s} = V$$

A decision must be made.
  a. Decrease pipe size to increase velocity - but Q will go down-is that acceptable in this system?
  b. Maintain pipe size and increase flow - velocity will go up, but so will HL. -is that acceptable?
  c. Get a smoother pipe
A compromise is necessary.

---

17. **Answer**   21.6 ft.

Without head loss in the pipe, the water level in B would be the same as A. Imagine Res. B as a large, wide piezometer; its water level will register the pressure there, in feet. Therefore the difference in the water levels in the two reservoirs - is the HL.

$$Q = .435 \ C \ d^{2.63} \ s^{.54}$$

$$1 = .435 \ 100 \ .5^{2.63} \ s^{.54}$$

$$.027 = s$$

HL = .027 ft./ft. x 800 ft. = 21.6 ft. difference in elevation of the water levels.

18. **Answer**   170 ft.

Change psi to feet:

> Point 1:   65 psi = 150.15 ft.
>
> Point 2:   28 psi = 64.68 ft.
>
> Elevation difference: 200 ft. - 115 ft. = 85 ft.

Set up Bernoulli's formula: Velocity Heads cancel.

> $Ph_1 + Z_1 = Ph_2 + Z_2 + HL$
>
> $150.15 + 200 = 64.68 + 115 + HL$
>
> 170.5 ft. = HL

The problem could also have been solved using Hazen-Williams formula.

> $Q = .435 \ C \ d^{2.63} \ s^{.54}$
>
> $1 = .435 \ 90 \ .5^{2.63} \ s^{.54}$
>
> $.0329 = s$
>
> HL = .0329 ft./ft. x 5200 ft. = 170 ft.

---

19. **Answer**   3 ft./sec.

Solve for diameter; solve for velocity.

> $Q = .435 \ C \ d^{2.63} \ s^{.54}$
>
> $5.28 = .435 \ 100 \ d^{2.63} \ .003^{.54}$
>
> 1.48 ft. = d
>
> 18 in. = d
>
> $Q = A \ V$
>
> $5.28 = .785 \times 1.5^2 \ V$
>
> 3 ft./sec. = V

---

20. **Answer**   2.94 cfs flow, 8.3 ft/sec. velocity

Draw diagram.   Calculate slope, solve for flow, velocity.

Available head   325 ft.

>   - 100   ft. subdivision elev.
>
>   225   ft.
>
>   - 46.2   ft. needed at hydrant
>
>   178.8   ft. potential head loss

1.5 miles x 5280 ft./mile = 7920 ft.

$$\text{Slope} = \frac{\text{HL}}{\text{L}} = \frac{178.8 \text{ ft.}}{7920 \text{ ft.}} = .0226 \text{ ft./ft.}$$

New PVC pipe C = 150

$Q = .435 \ C \ d^{2.63} \ s^{.54}$

$Q = .435 \ 150 \ .67^{2.63} \ .0226^{.54}$

$Q = 2.94 \text{ cfs}$

$Q = A \ V$

$2.94 = .789 \times .67^2 \ V$

$8.3 \text{ ft./sec.} = V$

---

21. **Answer**   The gages register no head loss - therefore, no flow.

---

22. **Answer**   A. 23.4 ft.
                  B. 5.8 ft.

$1.7 \text{ MGD} \times 1.55 \text{ cfs/MGD} = 2.64 \text{ cfs}$

A. $Q = .435 \ C \ d^{2.63} \ s^{.54}$

$2.64 = .435 \ 110 \ 1 \ s^{.54}$

$.0047 = s$

$\text{HL} = .0047 \text{ ft./ft.} \times 5000 \text{ ft.} = 23.4 \text{ ft.}$

B.   Insert the 16 inch line; solve for HL.

$Q = .435 \ C \ d^{2.63} \ s^{.54}$

$2.64 = .435 \ 110 \ 1.33^{2.63} \ s^{.54}$

$.0012 = s$

$\text{HL} = .0012 \text{ ft./ft.} \times 5000 \text{ ft.} = 5.8 \text{ ft. HL}$

---

23. **Answer**   97 psi

Change pressure to feet; subtract ft. of HL; change back to psi.

$400 \text{ psi} \times 2.31 \text{ ft./psi} = 924 \text{ ft.}$

$924 \text{ ft.} - 700 \text{ ft. HL} = 224 \text{ ft. remaining}$

$$\frac{224 \text{ ft.}}{2.31 \text{ ft./psi}} = 97 \text{ psi}$$

24. **Answer**   A. 20.6 ft.
                  B. C=150
                  C. 45.9 psi

A.   Volume $= .785 \times d^2 \times h \times 7.48$ gal./cu.ft.

                $200{,}000 = .785 \times d^2 \times 80 \times 7.48$

                20.6 ft. $= d$

B. Change gpm to cfs; solve for C.

$$\frac{700 \text{ gpm}}{60 \times 7.48} = 1.56 \text{ cfs}$$

$$Q = .435 \ C \ d^{2.63} \ s^{.54}$$

$$1.56 = .435 \ C \ 1 \ .001^{.54}$$

$$150 = C$$

C. Set up Bernoulli's formula;
     HL $= .5$ ft./1000 for a length of 4000 ft. $= 2$ ft. HL; Velocity Heads cancel.

$$Ph_1 + Z_1 = Ph_2 + Z_2 + HL$$

$$60 + 620 = Ph_2 + 570 + 4$$

$$106 \text{ ft.} = Ph_2$$

$$45.9 \text{ psi} = \text{pressure}$$

---

25. **Answer**   A. 6.9 cfs
                  B. 14.2 cfs

35 psi $= 80.85$ ft. Since head is 0 at downstream end, this is the HL.
$80.85/2000 = .04 = s$
Plastic C=150
14 in. $= 1.167$ ft.
13 in. $= 1.083$ ft.

A. Old Pipe: $Q = .435 \ C \ d^{2.63} \ s^{.54}$

                 $Q = .435 \ 60 \ 1.167^{2.63} \ .04^{.54}$

                 $Q = 6.9 \text{ cfs}$

B. New Pipe: $Q = .435 \ C \ d^{2.63} \ s^{.54}$

                 $Q = .435 \ 150 \ 1.083^{2.63} \ .04^{.54}$

                 $Q = 14.2 \text{ cfs}$

chapter 7

1. **Answer** 48.2 ft.

The difference in water surface elevations is the HL. Solve for HL in each pipe - add. Use Bernoulli's formula to obtain Ph in tank B.

Narrow Pipe

$Q = .435 \ C \ d^{2.63} \ s^{.54}$

$1 = .435 \ 90 \ .5^{2.63} \ s^{.54}$

$.0329 = s$

HL = .0329 ft./ft. x 1000 ft. = 32.9 ft.

Wide Pipe

$Q = .435 \ C \ d^{2.63} \ s^{.54}$

$1 = .435 \ 100 \ .67^{2.63} \ s^{.54}$

$.0065 = s$

HL = .0065 ft./ft. x 6000 ft. = 38.9 ft.

32.9 ft. + 38.9 ft. = 71.8 ft. total HL in system

Set points: point 1 at bottom of tank A and point 2 at bottom of tank B

The difference in velocity heads at a flow of 1 cfs will be negligible.

$Ph_1 + Z_1 = Ph_2 + Z_2 + HL$

$20 + 100 = Ph_2 + 0 + 71.8$

$48.2 \ ft. = Ph_2$

This is the depth of water in tank B.

2. **Answer** 11.1 ft.

Draw Diagram. Obtain flow. Solve for slope, and HL in each segment; add.

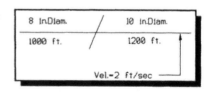

$Q = A \ V$

$Q = .785 \ x \ .83^2 \ x \ 2$

$Q = 1.1 \ cfs$

8 inch

$Q = .435 \ C \ d^{2.63} \ s^{.54}$

$1.1 = .435 \ 100 \ .67^{2.63} \ s^{.54}$

$.0078 = s$

HL = .0078 ft./ft. x 1000 ft. = 7.8 ft.

10 inch
$Q = .435 \ C \ d^{2.63} \ s^{.54}$

$1.1 = .435 \ 100 \ .83^{2.63} \ s^{.54}$

$.0027 = s$

HL = .0027 ft./ft. x 1200 ft. = 3.28 ft.

Adding: 7.8 ft. + 3.3 ft. = 11.1 ft. HL in whole system.

3. **Answer**   66.6 psi

Draw Diagram. Considered is the pressure change between two points, A & E; solve for elevation of E; find length of pipe A-E; solve for HL A-E; Use Bernoulli's formula to find pressure at E.

Pipe BC slopes up 4 ft./320 ft.

This equals 1.25 ft./100 ft. - to E.

Point E is at elevation 358.25 ft.

A-B (450 ft. of pipe)
$Q = .435 \ C \ d^{2.63} \ s^{.54}$

$.5 = .435 \ 100 \ .5^{2.63} \ s^{.54}$

$.0075 = s$

HL = .0075 ft./ft. x 450 ft. = 3.37 ft.

B-E (100 ft. of pipe)
$Q = .435 \ C \ d^{2.63} \ s^{.54}$

$.5 = .435 \ 100 \ .67^{2.63} \ s^{.54}$

$.0018 = s$

HL = .0018 ft./ft. x 100 ft. = .18 ft.

Adding: 3.37 ft. + .18 ft. = 3.55 ft. total HL

Vh difference insignificant - ignore.

$Ph_A + Z_A = Ph_E + Z_E + HL$

$168.63 + 347 = Ph_E + 358.25 + 3.55$

$153.83 \text{ ft.} = Ph_E$

$66.6 \text{ psi} = \text{pressure}$

4. **Answer**   A. 128750 ft.
   B. 1907 ft.

Draw Diagram.

10 in. Diam.	8 in.Diam.	6 in.Diam.
1400 ft.	1200 ft.	1500 ft.

A. Assume a flow; solve for s & HL in each pipe. Add. Choose equiv. pipe; solve for slope, then L.

Assume $Q = 1$ cfs

10 inch
$Q = .435 \ C \ d^{2.63} \ s^{.54}$

$1 = .435 \ \ 100 \ \ .83^{2.63} \ s^{.54}$

$.0023 = s$

$HL = .0023 \ \text{ft./ft.} \times 1400 \ \text{ft.} = 3.2 \ \text{ft.}$

8 inch
$Q = .435 \ C \ d^{2.63} \ s^{.54}$

$1 = .435 \ \ 100 \ \ .67^{2.63} \ s^{.54}$

$.0065 = s$

$HL = .0065 \ \text{ft./ft.} \times 1200 \ \text{ft.} = 7.8 \ \text{ft.}$

6 inch
$Q = .435 \ C \ d^{2.63} \ s^{.54}$

$1 = .435 \ \ 100 \ \ .5^{2.63} \ s^{.54}$

$.0270 = s$

$HL = .0270 \ \text{ft./ft.} \times 1500 \ \text{ft.} = 40.5 \ \text{ft.}$

Adding: 3.2 ft. + 7.8 ft. + 40.5 ft. = 51.5 ft. total HL

Equivalent 14 inch pipe; C=100
$Q = .435 \ C \ d^{2.63} \ s^{.54}$

$1 = .435 \ \ 100 \ \ 1.167^{2.63} \ s^{.54}$

$.0004 = s$

$s = \dfrac{HL}{L}$

$.0004 = \dfrac{51.5}{L}$

$128750 \ \text{ft.} = L$

B.   **Equivalent 6 inch pipe; C=100**
Slope already calculated at Q = 1 cfs; s = .0270.

$$s = \frac{HL}{L}$$

$$.0270 = \frac{51.5}{L}$$

L = 1907 ft.

5. **Answer**   8.6 ft.

Draw Diagram.   Newly constructed system;
change MGD to cfs; solve for HL in each
section; add.

1 MGD x 1.55 cfs/MGD = 1.55 cfs

Old
$$Q = .435 \ C \ d^{2.63} \ s^{.54}$$

$$1.55 = .435 \ 110 \ 1 \ s^{.54}$$

$$.0017 = s$$

HL = .0017 ft./ft. x 3000 ft. = 5.23 ft.

New
$$Q = .435 \ C \ d^{2.63} \ s^{.54}$$

$$1.55 = .435 \ 140 \ 1 \ s^{.54}$$

$$.0012 = s$$

HL = .0012 ft./ft. x 3000 ft. = 3.35 ft.

Adding: 5.23 ft. + 3.35 ft. = 8.58 ft. Total HL in system.

6. **Answer**   9 psi

Draw Diagram.   Solve for slope, head loss in
each pipe; add losses; change to psi.

8 inch
$$Q = .435 \ C \ d^{2.63} \ s^{.54}$$

$$2.3 = .435 \ 150 \ .67^{2.63} \ s^{.54}$$

$$.0143 = s$$

HL = .0143 ft./ft. x 1000 ft. = 14.3 ft.

4 inch

$$Q = .435 \ C \ d^{2.63} \ s^{.54}$$

$$2.3 = .435 \ 150 \ .33^{2.63} \ s^{.54}$$

$$.4514 = s$$

$$HL = .4514 \ \text{ft./ft.} \times 1000 \ \text{ft.} = 451 \ \text{ft.}$$

$$
\begin{array}{r}
14.3 \ \text{ft. HL 8 inch pipe} \\
+ \ 451.0 \ \text{ft. HL 4 inch pipe} \\
\hline
465.3 \ \text{ft Total HL}
\end{array}
$$

$$\frac{465.3 \ \text{ft.}}{2.31 \ \text{ft./psi}} = 201 \ \text{psi drop across the line}$$

$$
\begin{array}{r}
210 \ \text{psi upstream} \\
- \ 201 \ \text{psi Loss} \\
\hline
9 \ \text{psi downstream}
\end{array}
$$

7. **Answer**   198.5 ft.

Draw Diagram. Change MGD to cfs; miles to ft.; solve for slope, head loss in each pipe; add head losses.

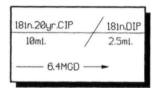

2.5 miles x 5280 ft./mile =

20 yr. old CIP C = 90

New DIP C = 140

3.4 MGD x 1.55 cfs/MGD = 5.27 cfs

18 inch CIP

$$Q = .435 \ C \ d^{2.63} \ s^{.54}$$

$$5.27 = .435 \ 90 \ 1.5^{2.63} \ s^{.54}$$

$$.0034 = s$$

$$.0034 \ \text{ft./ft.} \times 52800 \ \text{ft.} = 178.7 \ \text{ft. HL}$$

18 inch DIP

$$Q = .435 \ C \ d^{2.63} \ s^{.54}$$

$$5.27 = .435 \ 140 \ 1.5^{2.63} \ s^{.54}$$

$$.0015 = s$$

$$.0015 \ \text{ft./ft.} \times 13200 \ \text{ft.} = 19.8 \ \text{ft. HL}$$

$$
\begin{array}{lr}
\text{Adding head losses:} & 178.7 \ \text{ft.} \\
& \underline{19.8 \ \text{ft.}} \\
& 198.5 \ \text{ft. Total Head Loss}
\end{array}
$$

8.   **Answer**   151.8 ft.

Draw Diagram.   Change gpm to cfs; solve for
slope, head loss in each pipe; add head losses.

| 3000ft. | 7500ft. | 4080ft. |
| 14in. | 10in. | 6in. |

——685gpm——▶

unlined steel C = 140

$$\frac{685 \text{ gpm}}{7.48 \text{ gal./cu.ft. x } 60 \text{ sec./min.}} = 1.53 \text{ cfs}$$

14 inch
$Q = .435 \; C \; d^{2.63} \; s^{.54}$

$1.53 = .435 \; 140 \; 1.167^{2.63} \; s^{.54}$

$.0005 = s$

.0005 ft./ft. x 3000 ft. = 1.5 ft. HL

10 inch
$Q = .435 \; C \; d^{2.63} \; s^{.54}$

$1.53 = .435 \; 140 \; .83^{2.63} \; s^{.54}$

$.0027 = s$

.0027 ft./ft. x 7500 ft. = 20.3 ft. HL

6 inch
$Q = .435 \; C \; d^{2.63} \; s^{.54}$

$1.53 = .435 \; 140 \; .5^{2.63} \; s^{.54}$

$.0319 = s$

.0319 ft./ft. x 4080 ft. = 130 ft. HL

Adding Head Losses: 1.5 ft.
                       20.3 ft.
                 + 130.0 ft.
                    151.8 ft. Total HL

9.   **Answer**   .62 cfs

Draw Diagram.   Parallel pipe system.   Assume HL,
calculate flow through each pipe, get % split. apply to
2 cfs actual incoming flow to obtain recirculating flow.

Assume HL = 10 ft.

10 inch
Slope $= \dfrac{HL}{L} = \dfrac{10 \text{ ft.}}{1600 \text{ ft.}} = .0063$ ft./ft.

$$Q = .435 \ C \ d^{2.63} \ s^{.54}$$

$$Q = .435 \ 140 \ .83^{2.63} \ .0063^{.54}$$

$$Q = 2.42 \text{ cfs}$$

3 inch

$$\text{Slope} = \frac{HL}{L} = \frac{10 \text{ ft.}}{20 \text{ ft.}} = .0063 \text{ ft./ft.}$$

$$Q = .435 \ C \ d^{2.63} \ s^{.54}$$

$$Q = .435 \ 140 \ .25^{2.63} \ .6878^{.54}$$

$$Q = 1.1 \text{ cfs}$$

$$\begin{array}{r}
2.42 \text{ cfs (10 inch)} \\
+ \ 1.1 \text{ cfs (3 inch)} \\
\hline
3.52 \text{ cfs Total Flow (assumed)}
\end{array}$$

$$\frac{2.42 \text{ cfs}}{3.52 \text{ cfs}} = .69 = 69\% \text{ of flow through 10 inch}$$

$$\frac{1.1}{3.52} = .31 = 31\% \text{ of flow through 3 inch}$$

2 cfs (actual total flow) x .69 = 1.38 cfs through 10 inch

2 cfs (actual totoal flow) x .31 = .62 cfs through 3 inch

---

10. **Answer**   193 ft.

Draw Diagram.   Obtain head loss in 18 inch section. All remaining pressure is dissipated in 10" sections. Subtract upstream head loss from original pressure.

18 inch

$$Q = .435 \ C \ d^{2.63} \ s^{.54}$$

$$29 = .435 \ 140 \ 1.5^{2.63} \ s^{.54}$$

$$.035 = s$$

.035 ft./ft. x 200 ft. = 7 ft. HL

$$\begin{array}{r}
200 \text{ ft. pumping head} \\
- \ \ 7 \text{ ft. head loss in 18 inch pipe} \\
\hline
193 \text{ ft. head loss in either, or entire 10 inch section}
\end{array}$$

11. **Answer**   A. 93 ft.
                  B. 70.3 ft.
                  C. 70.3 ft.

A. Entire flow goes through bottom loop; solve for s, HL.

$$Q = .435 \ C \ d^{2.63} \ s^{.54}$$

$$10 = .435 \ \ 120 \ \ .83^{2.63} \ \ s^{.54}$$

$$.1162 = s$$

$$HL = .1162 \ ft./ft. \ x \ 800 \ ft. \ = 93 \ ft.$$

B. Flow splits; assume HL; solve for slope, then flow in each; add flows for total flow; get % flow split; apply to real flow; using real flow, solve for slope, HL in bottom loop.

Assume HL = 10

$$s = \frac{HL}{L} = \frac{10}{1200} = .0083 \ ft./ft.$$

6 inch
$$Q = .435 \ C \ d^{2.63} \ s^{.54}$$

$$Q = .435 \ \ 90 \ \ .5^{2.63} \ \ .0083^{.54}$$

$$Q = .48 \ cfs$$

$$s = \frac{HL}{L} = \frac{10}{800} = .0125$$

10 inch
$$Q = .435 \ C \ d^{2.63} \ s^{.54}$$

$$Q = .435 \ \ 120 \ \ .83^{2.63} \ \ .0125^{.54}$$

$$Q = 3.0 \ cfs$$

Adding flows: .48 cfs + 3.0 cfs = 3.48 cfs total flow (from assumed HL)

$$\frac{.48 \ cfs}{3.48 \ cfs} = .14 \quad \text{Flow through top loop is 14\% of total flow.}$$

$$\frac{3 \ cfs}{3.48 \ cfs} = .86 \quad \text{Flow through bottom loop is 86\% of total flow.}$$

10 cfs (actual total flow) x .14 = 1.4 cfs (actual flow through top loop)
10 cfs (actual total flow) x .86 = 8.6 cfs (actual flow through bottom loop)

Solve for actual HL in bottom loop.
$$Q = .435 \ C \ d^{2.63} \ s^{.54}$$

$$8.6 = .435 \ \ 120 \ \ .83^{2.63} \ \ s^{.54}$$

$$.0879 = s$$

$$HL = .0879 \ ft./ft. \ x \ 800 \ ft. \ = 70.3 \ ft.$$

C. HL across the entire system should be the same. We will work it out to prove it.

Convert to equivalent pipe for entire system; in section b we assumed a HL of 10 ft., and produced a total flow based on this of 3.48 cfs. Now with this flow solve for slope in equivalent pipe, then solve for length of it; then put actual flow through this pipe to get HL.

Equivalent pipe: Choose 12 inch diam. pipe C=100

$$Q = .435 \ C \ d^{2.63} \ s^{.54}$$

$$3.48 = .435 \ 100 \ 1 \ s^{.54}$$

$$.0093 = s$$

$$s = \frac{HL}{L}$$

$$.0093 = \frac{10}{L}$$

1075 ft. = L (of equiv. pipe)

$$Q = .435 \ C \ d^{2.63} \ s^{.54}$$

$$10 = .435 \ 100 \ 1 \ s^{.54}$$

$$.0657 = s$$

HL = .0657 ft./ft. x 1075 ft. = 70.5 ft.   Right On!

12. **Answer**   8.4 ft.

Draw Diagram. Assume HL; solve for flow in each; add up; obtain % flow split; multiply by actual flow (given-total) to obtain flow through that section; solve for HL.

12 in Diam
1500 ft.
3 cfs
C-100 All Pipes
2000 ft.
6 in Diam

Assume HL = 20 ft.

$$s = \frac{HL}{L} = \frac{20}{1500} = .0133 \ \text{ft./ft.}$$

12 inch
$$Q = .435 \ C \ d^{2.63} \ s^{.54}$$

$$Q = .435 \ 100 \ 1 \ .0133^{.54}$$

$$Q = 4.2 \ \text{cfs}$$

$$s = \frac{HL}{L} = \frac{20}{2000} = .01 \ \text{ft./ft.}$$

6 inch

$$Q = .435 \ C \ d^{2.63} \ s^{.54}$$

$$Q = .435 \ 100 \ .5^{2.63} \ .01^{.54}$$

$$Q = .6 \ cfs$$

Adding flows: 4.2 cfs + .6 cfs = 4.8 cfs Total flow (from assumed HL)

$$\frac{4.2 \ cfs}{4.8 \ cfs} = .88 \quad \text{(top flow is 88\% of total flow)}$$

$$\frac{.6 \ cfs}{4.8 \ cfs} = .12 \quad \text{(bottom flow is 12\% of total flow)}$$

Apply top % to actual flow:

3 cfs x .88 = 2.64 cfs

Top
$$Q = .435 \ C \ d^{2.63} \ s^{.54}$$

$$2.64 = .435 \ 100 \ 1 \ s^{.54}$$

$$.0056 = s$$

HL = .0056 ft./ft. x 1500 ft. = 8.4 ft.

---

13. **Answer**   16.8 ft./sec.

Solve for flow through bottom section; subtract from total flow to get flow through top; solve for velocity in top.

$$Q = A \ V$$

$$Q = .785 \ x \ .67^2 \ x \ 2$$

$$Q = .7 \ cfs \ (bottom)$$

4 cfs total flow - .7 cfs bottom flow = 3.3 cfs top

$$Q = A \ V$$

$$3.3 = .785 \ x \ .5^2 \ V$$

$$16.8 \ ft./sec. = V$$

---

14. **Answer**   6 inches

Draw Diagram. Convert psi to ft. Solve for s & HL through 6 inch pipe. Subtract from total HL for HL in unknown. Solve for s, then diameter.

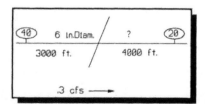

20 psi x 2.31 ft./psi = 46.2 ft. total HL in system

6 inch

$$Q = .435 \; C \; d^{2.63} \; s^{.54}$$

$$.3 = .435 \; 80 \; .5^{2.63} \; s^{.54}$$

$$.0044 = s$$

HL = .0044 ft./ft. x 3000 ft. = 13.2 ft. HL in 6 inch section.

46.2 ft. Tot. HL in system - 13.2 ft. HL in 6 inch = 33.0 ft. HL in unknown.

$$s = \frac{HL}{L} = \frac{33}{4000} = .0083 \text{ ft./ft. slope of unknown}$$

Unknown

$$Q = .435 \; C \; d^{2.63} \; s^{.54}$$

$$.3 = .435 \; 80 \; d^{2.63} \; .0083^{.54}$$

$$.44 \text{ ft.} = d$$

Assume 6 inch diameter pipe.

---

15. **Answer**   60.9 cfs

Draw Diagram. Elevation difference = the total HL. Assume a flow; solve for HL in each pipe of actual system; add HL's; Create equivalent pipe; with assumed flow, solve for slope; using this and calculated total HL, solve for length of equiv. pipe. Insert again with actual head loss, solving for slope; solve for flow through equiv. pipe. This is the actual flow through the system.

700 ft. - 460 ft. = 240 ft. = HL in system.

Assume flow = 22 cfs

24 inch

$$Q = .435 \; C \; d^{2.63} \; s^{.54}$$

$$22 = .435 \; 130 \; 2^{2.63} \; s^{.54}$$

$$.006 = s$$

HL = .006 ft./ft. x 1300 = 7.8 ft.

20 inch

$$Q = .435 \; C \; d^{2.63} \; s^{.54}$$

$$22 = .435 \; 130 \; 1.67^{2.63} \; s^{.54}$$

$$.0143 = s$$

HL = .0143 ft./ft. x 2000 ft. = 28.6 ft.

Adding HL's: 7.8 ft. + 28.6 ft. = 36.4 ft. total HL (from assumed flow)

Create equivalent pipe: 12 inch diam. C=100; Solve for slope.

$Q = .435 \ C \ d^{2.63} \ s^{.54}$

$22 = .435 \ 100 \ 1 \ s^{.54}$

$.283 = s$

$s = .\dfrac{HL}{L}$

$.283 = \dfrac{36.4}{L}$

$128.6 = L$ (equiv. pipe length)

$s = \dfrac{HL}{L} = \dfrac{240}{128.6} = 1.866 \ \text{ft./ft.}$

$Q = .435 \ C \ d^{2.63} \ s^{.54}$

$Q = .435 \ 100 \ 1 \ 1.866^{.54}$

$Q = 60.9 \ \text{cfs}$

---

16. **Answer**   System B

Assume any flow; calculate for total HL in each system; the one with the greater potential HL, will carry less flow.

System A: Choose a flow; solve for HL in each segment; add up.

$Q = 2 \ \text{cfs}$

16 inch
$Q = .435 \ C \ d^{2.63} \ s^{.54}$

$2 = .435 \ 120 \ 1.33^{2.63} \ s^{.54}$

$.0006 = s$

HL = .0006 ft./ft. x 9000 ft. = 5.3 ft.

12 inch
$Q = .435 \ C \ d^{2.63} \ s^{.54}$

$2 = .435 \ 120 \ 1 \ s^{.54}$

$.0024 = s$

HL = .0024 ft./ft. x 6000 ft. = 14.3 ft.

10 inch

$$Q = .435 \; C \; d^{2.63} \; s^{.54}$$

$$2 = .435 \;\; 120 \;\; .83^{2.63} \; s^{.54}$$

$$.0059 = s$$

$$HL = .0059 \; ft./ft. \times 3000 \; ft. = 17.7 \; ft.$$

Adding: 5.3 ft. + 14.3 ft. + 17.7 ft. = 37.3 ft. total HL

System B: Work on parallel segment first, assume a HL; solve for flow in top and bottom segments; add flows; choose equivalent pipe; using new total flow, solve for slope, then length. Then treat as a series pipe system with equivalent pipe in middle.

Parallel segment: assume HL = 10 ft.

Top

$$s = \frac{HL}{L} = \frac{10}{5000} = .002 \; ft./ft.$$

$$Q = .435 \; C \; d^{2.63} \; s^{.54}$$

$$Q = .435 \;\; 120 \;\; .67^{2.63} \;\; .002^{.54}$$

$$Q = .635 \; cfs$$

Bottom

$$s = \frac{HL}{L} = \frac{10}{7000} = .0014$$

$$Q = .435 \; C \; d^{2.63} \; s^{.54}$$

$$Q = .435 \;\; 120 \;\; .83^{2.63} \;\; .0014^{.54}$$

$$Q = .92 \; cfs$$

Adding: (based on assumed HL)

.635 cfs + .92 cfs = 1.56 cfs total flow (from assumed HL)

Equivalent pipe:

$$Q = .435 \; C \; d^{2.63} \; s^{.54}$$

$$1.56 = .435 \;\; 120 \;\; .83^{2.63} \; s^{.54}$$

$$.0037 = s$$

Solve for Length.

$$s = \frac{HL}{L}$$

$$.0037 = \frac{10}{L}$$

$$2703 \; ft. = L$$

Now there are three series segments; solve for HL in each, then add.

18 inch

$Q = .435 \ C \ d^{2.63} \ s^{.54}$

$2 = .435 \ 120 \ 1.5^{2.63} \ s^{.54}$

$.0003 = s$

$HL = .0003 \ \text{ft./ft.} \times 11,000 \ \text{ft.} = 3.3 \ \text{ft.}$

Equiv.

$Q = .435 \ C \ d^{2.63} \ s^{.54}$

$2 = .435 \ 120 \ .83^{2.63} \ s^{.54}$

$.0059 = s$

$HL = .0059 \ \text{ft./ft.} \times 2703 \ \text{ft.} = 15.9 \ \text{ft.}$

10 inch

$Q = .435 \ C \ d^{2.63} \ s^{.54}$

$2 = .435 \ 120 \ .83^{2.63} \ s^{.54}$

$.0059 = s$

$HL = .0059 \ \text{ft./ft.} \times 2500 \ \text{ft.} = 14.8 \ \text{ft.}$

Adding: 3.3 ft. + 15.9 ft. + 14.8 ft. = 34 ft. total HL

---

17. **Answer** 17.9 ft.

Elevation difference = total HL in pipe system.
Section 3: velocity given - solve for flow.
Section 2 & 3: find slope & HL; add for total HL bottom.
Section 4: solve for slope; find flow.
Add flows in section 4 and 2-3; this gives incoming or outgoing flow.
Section 5: find slope and HL
Add HL's section 1,4,5 for total HL which equals the elevation difference.

$Q = A \ V$

$Q = .785 \times .83^2 \times 2$

$Q = 1.1$

Section 3

$Q = .435 \ C \ d^{2.63} \ s^{.54}$

$1.1 = .435 \ 100 \ .83^{2.63} \ s^{.54}$

$.0027 = s$

$HL = .0027 \ \text{ft./ft.} \times 1500 \ \text{ft.} = 4.1 \ \text{ft.}$

Section 2

$Q = .435 \ C \ d^{2.63} \ s^{.54}$

$1.1 = .435 \ 100 \ .67^{2.63} \ s^{.54}$

$.0078 = s$

$HL = .0078 \ \text{ft./ft.} \ x \ 800 \ \text{ft.} = 6.2 \ \text{ft.}$

Adding HL's: $4.1 \ \text{ft.} + 6.2 \ \text{ft.} = 10.3 \ \text{ft. total HL bottom}$

Therefore: 10.3 ft. is HL top also.

Section 4

$s = \dfrac{HL}{L} = \dfrac{10.3}{1000} = .0103 \ \text{ft./ft.}$

$Q = .435 \ C \ d^{2.63} \ s^{.54}$

$Q = .435 \ 100 \ .83^{2.63} \ .0103^{.54}$

$Q = 2.2 \ \text{cfs}$

Adding flows: $1.1 \ \text{cfs} + 2.2 \ \text{cfs} = 3.3 \ \text{cfs flow in 1 \& 5}$

Section 1 & 5
Combine length 1 & 5 = 900 ft.

$Q = .435 \ C \ d^{2.63} \ s^{.54}$

$3.3 = .435 \ 100 \ 1 \ s^{.54}$

$.0084 = s$

$HL = .0084 \ \text{ft./ft.} \ x \ 900 \ \text{ft.} = 7.6 \ \text{ft.}$

Add HL's: Top (4) 10.3 ft. + Ends (1 & 5) 7.6 ft. = 17.9 ft.

18. **Answer** A. 2950 ft.
 B. 91.8 ft.

Draw Diagram.

A. Assume a HL; obtain slope, flow
 for each pipe in actual system; add
 flows for total flow; create
 equivalent pipe; using total flow
 from assumed HL, solve for slope;
 with assumed HL & slope of equiv.
 pipe, solve for length.

Assume HL = 100 ft.

4 inch

$$s = \frac{HL}{L} = \frac{100}{570} = .1754 \text{ ft.}$$

$$Q = .435 \; C \; d^{2.63} \; s^{.54}$$

$$Q = .435 \; 140 \; .33^{2.63} \; .1754^{.54}$$

$$Q = 1.3 \text{ cfs}$$

6 inch

$$s = \frac{HL}{L} = \frac{100}{900} = .1111 \text{ ft.}$$

$$Q = .435 \; C \; d^{2.63} \; s^{.54}$$

$$Q = .435 \; 80 \; .5^{2.63} \; .1111^{.54}$$

$$Q = 1.7 \text{ cfs}$$

8 inch

$$s = \frac{HL}{L} = \frac{100}{1200} = .0833 \text{ ft.}$$

$$Q = .435 \; C \; d^{2.63} \; s^{.54}$$

$$Q = .435 \; 100 \; .67^{2.63} \; .0833^{.54}$$

$$Q = 4.0 \text{ cfs}$$

Adding flows: (based on assumed HL)

1.3 cfs + 1.7 cfs + 4.0 cfs = 7.0 cfs Total flow (from assmed HL)

Choose equivalent pipe: 12 inch diam. C=100

$$Q = .435 \; C \; d^{2.63} \; s^{.54}$$

$$7 = .435 \; 100 \; 1 \; s^{.54}$$

$$.0339 = s$$

Solve for length of equivalent pipe.

$$s = \frac{HL}{L}$$

$$.0339 = \frac{100}{L}$$

$$2950 \text{ ft.} = L$$

B.   Use actual flow to obtain HL of equivalent pipe (which is the same as the HL of our system).

$$\frac{3000 \text{ gpm}}{7.48 \times 60} = 6.68 \text{ cfs}$$

$$Q = .435 \ C \ d^{2.63} \ s^{.54}$$

$$6.68 = .435 \ 100 \ 1 \ s^{.54}$$

$$.0311 = s$$

$$HL = .0311 \ ft./ft. \times 2950 \ ft. = 91.8 \ ft.$$

19. **Answer**   A. 5061 ft.
               B. 6.75 cfs

Draw Diagram. Assume Q = 1.

20 in.Diam.	16 in.Diam.	12 in.Diam.
10,000 ft.	8000 ft.	L - ?
C-120	C-120	C-120

A. Solve for HL in 15 inch pipe. Total HL in actual system is the same. Using assumed flow, obtain HL for each of the first two pipes. Difference between HL (equiv. pipe) and HL in actual system just calculated, is HL in the 12 inch pipe - at that flow. Solve for length of 12 inch pipe. This will be the correct pipe length to yield a HL that makes the actual system equiv. to the equiv. pipe.

Equiv.
$$Q = .435 \ C \ d^{2.63} \ s^{.54}$$

$$1 = .435 \ 100 \ 1.25^{2.63} \ s^{.54}$$

$$.0003 = s$$

$$HL = .0003 \ ft./ft. \times 16500 \ ft. = 5.14 \ ft.$$

20 inch
$$Q = .435 \ C \ d^{2.63} \ s^{.54}$$

$$Q = .435 \ 120 \ 1.67^{2.63} \ s^{.54}$$

$$.00005 = s$$

$$HL = .00005 \ ft./ft. \times 10000 \ ft. = .5 \ ft.$$

16 inch
$$Q = .435 \ C \ d^{2.63} \ s^{.54}$$

$$Q = .435 \ 120 \ 1.33^{2.63} \ s^{.54}$$

$$.00016 = s$$

$$HL = .00016 \ ft./ft. \times 8000 \ ft. = 1.3 \ ft.$$

Adding:
.5 ft. + 1.3 ft. = 1.8 ft. HL in AB & BC of actual system at Q=1

5.14 ft. (HL Equiv. pipe at Q=1) - 1.8 ft. = 3.34 ft. (HL CD at Q=1)

12 inch

$$Q = .435 \ C \ d^{2.63} \ s^{.54}$$

$$1 = .435 \ 120 \ 1 \ s^{.54}$$

$$.00066 = s$$

$$s = \frac{HL}{L}$$

$$.00066 = \frac{3.34}{L}$$

$$5061 \ \text{ft.} = L$$

B.   Tot. HL is given; to solve for flow we must use the equivalent pipe method. We have already worked through the procedure in section A. Use the same figures. Assume a flow (1); solve for HL in each pipe of actual system; add, solve for slope in equiv. pipe (use the same one); find a new length for it; Using s=HL/L, insert new L and 135 ft. for HL, and get s; solve for Q. This is the flow through the actual system.

Assume Q = 1 cfs.

From section A:
HL in 20 inch pipe = .5 ft.

HL in 16 inch pipe = 1.3 ft.

s for 12 inch pipe = .00066

HL (12 inch) = .00066 ft./ft. x 3000 ft. = 1.98 ft.

Adding: .5 ft. + 1.3 ft. + 1.98 ft. = 3.78 ft. total HL at Q=1

From section A:
s for equiv. pipe = .0003 ft./ft.

Solve for length:

$$s = \frac{HL}{L}$$

$$.0003 = \frac{3.78}{L}$$

$$12600 \ \text{ft.} = L$$

Now get real slope for equiv. pipe.

$$s = \frac{HL}{L} = \frac{135}{12600} = .0107 \ \text{ft./ft.}$$

Solve for flow in equiv. pipe.

$$Q = .435 \ C \ d^{2.63} \ s^{.54}$$

$$Q = .435 \ 100 \ 1.25^{2.63} \ .0107^{.54}$$

$$Q = 6.75 \ \text{cfs}$$

20. **Answer** .175 cfs

Draw Diagram. HL across dishwasher and sink lines are equal. Solve for flow in each; add. This is the flow approaching the split. Solve for flow in shower line. Add to downstream flow. This is flow entering house.

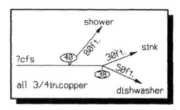

To sink:
38 psi x 2.31 ft./psi = 87.8 ft.

$$\frac{87.8 \text{ ft. HL}}{30 \text{ ft. pipe}} = 2.926 \text{ ft./ft. slope}$$

$Q = .435 \ C \ d^{2.63} \ s^{.54}$

$Q = .435 \ 140 \ .0625^{2.63} \ 2.926^{.54}$

$Q = .074$ cfs

To dishwasher:
38 psi x 2.31 ft./psi = 87.8 ft.

$$\frac{87.8 \text{ ft. HL}}{50 \text{ ft. pipe}} = 1.756 \text{ ft./ft slope}$$

$Q = .435 \ C \ d^{2.63} \ s^{.54}$

$Q = .435 \ 140 \ .0625^{2.63} \ 1.756^{.54}$

$Q = .056$ cfs

    .074 cfs to sink
  - .056 cfs to dishwasher
    .13 cfs approaching split.

To shower:
40 psi x 2.31 ft./psi = 92.4 ft.

$$\frac{92.4 \text{ ft. HL}}{80 \text{ ft. pipe}} = 1.155 \text{ ft./ft. slope}$$

$Q = .435 \ C \ d^{2.63} \ s^{.54}$

$Q = .435 \ 140 \ .0625^{2.63} \ 1.155^{.54}$

$Q = .045$ cfs

    .13 cfs approaching split
  - .045 cfs to shower
    .175 cfs entering house

21. **Answer** 6.7 cfs

HL is given as the difference in elevation between the two reservoirs (30 ft.). Bottom segment of parallel system is composed of two pipes in series. Convert to equivalent pipe: assume flow, find HL in each, add, find slope of equivalent pipe, solve for length. Parallel section is now composed of two pipes; convert to one equiv. pipe: assume HL, find flow in each, add; using total flow, find slope for equivalent; solve for length of equivalent pipe. Entire system now composed of 3 pipes in series; convert to one equivalent pipe: assume flow, find HL in each, add; using assumed flow, find slope of equivalent, solve for length. Now take the actual HL (30 ft.) and solve for s and flow for entire system.

Bottom of Parallel System
Assume Q = 10 cfs   Using actual system:

Actual 16 inch
$Q = .435 \ C \ d^{2.63} \ s^{.54}$

$10 = .435 \ 120 \ 1.33^{2.63} \ s^{.54}$

$.0117 = s$

HL = .0117 ft./ft. x 6000 ft. = 70.1 ft.

12 inch
$Q = .435 \ C \ d^{2.63} \ s^{.54}$

$10 = .435 \ 120 \ 1 \ s^{.54}$

$.0469 = s$

HL = .0469 ft./ft. x 6000 ft. = 281.3 ft.

Adding HL's:
70.1 ft. + 281.3 ft. = 351.4 ft. Total HL (from assumed flow).

Equivalent pipe - choose 12 inch diam. C=120

From above, slope at this flow is .0469.

$s = \dfrac{HL}{L}$

$.0469 = \dfrac{351.4}{L}$

7493 ft. = L

The pipe now becomes the bottom segment of a parallel system. (7493 ft. of 12 inch pipe C=120).

Convert parallel system to equivalent pipe:

Assume HL = 20 ft. Solve for flow.

Actual 20 inch

$$s = \frac{HL}{L} = \frac{20}{8000} = .0025 \text{ ft./ft.}$$

$$Q = .435 \ C \ d^{2.63} \ s^{.54}$$

$$Q = .435 \ 120 \ 1.67^{2.63} \ .0025^{.54}$$

$$Q = 7.9 \text{ cfs}$$

New Equiv. 12 inch

$$s = \frac{HL}{L} = \frac{20}{7493} = .0027 \text{ ft./ft.}$$

$$Q = .435 \ C \ d^{2.63} \ s^{.54}$$

$$Q = .435 \ 120 \ 1 \ .0027^{.54}$$

$$Q = 2.14 \text{ cfs}$$

Adding flows: 7.9 cfs + 2.14 cfs = 10 cfs

Choose equivalent pipe: 12 inch diam. C=120; solve for slope.

$$Q = .435 \ C \ d^{2.63} \ s^{.54}$$

$$10 = .435 \ 120 \ 1 \ s^{.54}$$

$$.0469 = s$$

$$s = \frac{HL}{L}$$

$$.0469 = \frac{20}{L} = 426 \text{ ft. length}$$

This pipe now becomes the middle segment of the series system. (426 ft. of 12 inch pipe C=120)

Convert this to equivalent pipe: Assume Q=1 cfs.

20 inch

$$Q = .435 \ C \ d^{2.63} \ s^{.54}$$

$$1 = .435 \ 100 \ 1.67^{2.63} \ s^{.54}$$

$$.00008 = s$$

$$HL = .00008 \text{ ft./ft. x 4000 ft. } = .3 \text{ ft.}$$

12 inch

$$Q = .435 \ C \ d^{2.63} \ s^{.54}$$

$$1 = .435 \ 120 \ 1 \ s^{.54}$$

$$.0007 = s$$

$$HL = .0007 \text{ ft./ft. x 426 ft. } = .3 \text{ ft.}$$

24 inch

$$Q = .435 \ C \ d^{2.63} \ s^{.54}$$

$$1 = .435 \ 100 \ 2^{2.63} \ s^{.54}$$

$$.00003 = s$$

$$HL = .00003 \ \text{ft./ft.} \times 3000 \ \text{ft.} = .1 \ \text{ft.}$$

Adding HL's:

.3 ft. + .3 ft. + .1 ft. = .7 ft. HL (based on assumed Q=1 cfs)

Choose equivalent pipe: 18 inch C=100; solve for slope.

$$Q = .435 \ C \ d^{2.63} \ s^{.54}$$

$$1 = .435 \ 100 \ 1.5^{2.63} \ s^{.54}$$

$$.0001 = s$$

$$s = \frac{HL}{L}$$

$$.0001 = \frac{.7}{L}$$

$$7000 \ \text{ft.} = L$$

The pipe now becomes one: (7000 ft. of 18 inch pipe C=100)

Use actual HL given; solve for slope

$$s = \frac{HL}{L} = \frac{30}{7000} = .0043 \ \text{ft./ft.}$$

$$Q = .435 \ C \ d^{2.63} \ s^{.54}$$

$$Q = .435 \ 100 \ 1.5^{2.63} \ .0043^{.54}$$

$$Q = 6.7 \ \text{cfs}$$

---

22. **Answer**   1.48 cfs

Bottom section is composed of three pipes in series. Convert to one equivalent pipe: assume flow; solve for HL in actual system; add HL's; create equivalent; find slope; solve for length. Now this is a parallel system; assume HL; solve for flow in each; add; get % split; apply to actual flow for flow in CD.

Convert bottom to equivalent pipe:

Assume Q=1 cfs.

8 inch

$Q = .435 \ C \ d^{2.63} \ s^{.54}$

$1 = .435 \ 120 \ .67^{2.63} \ s^{.54}$

$.0046 = s$

$HL = .0046 \ \text{ft./ft.} \times 700 \ \text{ft.} = 3.25 \ \text{ft.}$

6 inch

$Q = .435 \ C \ d^{2.63} \ s^{.54}$

$1 = .435 \ 100 \ .5^{2.63} \ s^{.54}$

$.027 = s$

$HL = .027 \ \text{ft./ft.} \times 500 \ \text{ft.} = 13.5 \ \text{ft.}$

8 inch

$Q = .435 \ C \ d^{2.63} \ s^{.54}$

$1 = .435 \ 110 \ .67^{2.63} \ s^{.54}$

$.0055 = s$

$HL = .0055 \ \text{ft./ft.} \times 1000 \ \text{ft.} = 5.5 \ \text{ft.}$

Adding HL's: 3.25 ft. + 13.5 ft. + 5.5 ft. = 22.3 ft. total HL

Choose equivalent pipe: 8 inch diam. C=110

From above, slope for this = .0055 ft./ft..

$s = \dfrac{HL}{L}$

$.0055 = \dfrac{22.3}{L}$

$4049 \ \text{ft.} = L$

Now a parallel system remains; solve for flows.

Assume an HL: (50 ft)

Top

$s = \dfrac{HL}{L} = \dfrac{50}{1500} = .0333 \ \text{ft./ft.}$

$Q = .435 \ C \ d^{2.63} \ s^{.54}$

$Q = .435 \ 130 \ .83^{2.63} \ .0333^{.54}$

$Q = 5.1 \ \text{cfs}$

Bottom

$s = \dfrac{HL}{L} = \dfrac{50}{4049} = .0124 \ \text{ft./ft.}$

$Q = .435 \ C \ d^{2.63} \ s^{.54}$

$Q = .435 \ 110 \ .67^{2.63} \ .0124^{.54}$

$Q = 1.6$ cfs

Adding flows:
5.1 cfs + 1.6 cfs = 6.7 cfs Total flow (from assumed HL)

$\dfrac{5.1}{6.7} = .76$ (flow through top is 76% of total flow)

$\dfrac{1.6}{6.7} = .24$ (flow through bottom is 24% of total flow)

Bottom section carries 24% of total flow.

6.18 cfs x .24 = 1.48 cfs

chapter 8

1. **Answer**   A. 13 ft. equivalent pipe
                  B. 10 ft. equivalent pipe
                  C. 6.5 ft. equivalent pipe
                  D. 90 ft. equivalent pipe
                  E. 6.5 ft. equivalent pipe

2. **Answer**   20 ft. equivalent pipe

This is a borda entrance.
Borda entrance (8 in.) = 20 ft.

3. **Answer**   .04 ft.

Draw Diagram. Change fitting to equivalent length of pipe; calculate slope, head loss.

ordinary entrance (10 in.) = 15 ft. equiv. pipe

galvanized iron pipe C = 120

$Q = .435 \ C \ d^{2.63} \ s^{.54}$

$1.3 = .435 \ 120 \ .83^{2.63} \ s^{.54}$

$.0027 = s$

.0027 ft./ft. x 15 ft. = .04 ft. HL generated by fitting

4. **Answer**   7.1 cfs drop in flow

Calculate flow through pipe without check valve.   Then adjust pipe length for fitting; recalculate slope, and flow.

Without Valve.
6.8 psi x 2.31 ft./psi = 15.7 ft. HL

$$\text{Slope} = \frac{HL}{length} = \frac{15.7 \text{ ft.}}{350 \text{ ft.}} = .0449 \text{ ft./ft.}$$

$Q = .435 \ C \ d^{2.63} \ s^{.54}$

$Q = .435 \ 140 \ 1.67^{2.63} \ .0449^{.54}$

Q = 43.9 cfs flowing, without valve.

With Valve.
pipe length   350 ft.
check valve (20 in.)   + 135 ft.
                          485 ft. equivalent length pipe

$$\text{Slope} = \frac{HL}{length} = \frac{15.7 \text{ ft.}}{485 \text{ ft.}} = .0324 \text{ ft./ft.}$$

$Q = .435 \ C \ d^{2.63} \ s^{.54}$

$Q = .435 \ 140 \ 1.67^{2.63} \ .0324^{.54}$

Q = 36.8 cfs flowing, with valve.

43.9 cfs - 36.8 cfs = 7.1 cfs drop in flow

---

5. **Answer**   14.9 psi

Draw Diagram.   Calculate equivalent pipe length; solve for slope, head loss; change to psi.

3 inch
actual pipe length   50 ft.
sudden enlargement 1/2   + 5 ft.
                                      55 ft.

$Q = .435 \ C \ d^{2.63} \ s^{.54}$

$.8 = .435 \ 100 \ .25^{2.63} \ s^{.54}$

$.5232 = s$

.5232 ft./ft. x 55 ft. = 28.8 ft. HL

6 inch

$$Q = .435 \ C \ d^{2.63} \ s^{.54}$$

$$.8 = .435 \ \ 100 \ \ .5^{2.63} \ s^{.54}$$

$$.1139 = s$$

.1139 ft./ft. x 50 ft. = 5.7 ft. HL

Adding head losses: 28.8 ft.
                                 + 5.7 ft.
                                 34.5 ft. total head loss

$$\frac{34.5 \ ft.}{2.31 \ ft./psi} = 14.9 \ psi$$

---

6. **Answer**   4 ft.

Obtain equivalent length for fittings; add; solve for HL.

glove valve (6 inches)	160 ft.
45 deg. elbow (2) (6 inches)	+ 16 ft.
	176 ft. length

$$Q = .435 \ C \ d^{2.63} \ s^{.54}$$

$$1 = .435 \ \ 110 \ \ .5^{2.63} \ s^{.54}$$

$$.0227 = s$$

HL = .0227 ft./ft. x 176 ft. = 4 ft. HL created by fittings

---

7. **Answer**   A. 20 ft.
                  B. 4.1 ft.

A. Obtain equivalent length of valve; add to pipe length; solve for HL.

3/4 closed gate valve (12 in.)	800 ft.
actual pipe length	+ 200 ft.
	1000 ft.

$$s = \frac{HL}{L}$$

$$.02 = \frac{HL}{1000 \ ft.}$$

20 ft. = HL

B. Repeat with fully open gate valve.

Fully open gate valve (12 in.)   6.5 ft.
Actual pipe length   + 200.0 ft.
                                        206.5 ft.

206.5 ft. x .02 ft./ft. = 4.1 ft. HL

---

8. **Answer**   40 ft.

Add straight pipe lengths; convert fittings to equiv. pipe; add on; solve for HL.

Straight pipe (80 ft. + 30 ft. + 50 ft.)   160   ft.
Borda entrance (6 in.)   18   ft.
Standard elbows (2) (6 in.)   34   ft.
Gate valve (6 in.)   3.5   ft.
Ordinary exit (6 in.)   + 18   ft.
                                           233.5   ft. total pipe length

10 yr. old CIP: C=110

$$Q = .435 \ C \ d^{2.63} \ s^{.54}$$

$$3 = .435 \ \ 110 \ \ .5^{2.63} \ s^{.54}$$

$$.1733 = s$$

HL = .1733 ft./ft. x 233.5 ft. = 40 ft.

Note: Tank #1 is only 20 ft. deep. The HL is 40 ft. If the 30 ft. length of pipe is a horizontal pipe, the water would not get to Tank #2 at this flow. In order to maintain this flow, there must be a 20 ft. vertical drop in this pipe.

---

9. **Answer**   .3%.

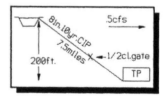

The head loss in the system is determined by slope and pipe length. Calculate percentage of pipe length that this valve equates to.

7.5 miles of pipe x 5280 ft./mi. = 39600 ft.

ordinary exit (8 in.)   19 ft.
half closed gate (8 in.)   130 ft.
ordinary entrance (8 in.)   12 ft.
actual pipe length   + 39600 ft.
                                     39761 ft.

$$\frac{130 \ \text{ft. equiv. pipe length for half closed gate}}{39761 \ \text{ft. total equiv. pipe length}} = .00327 = .3\%$$

10.   **Answer**   .095 cfs

Draw Diagram.   Solve for head loss; obtain equivalent
length for fittings; add to pipe length; solve for slope, then
flow.

New copper pipe (C=140)

Head loss = pressure at entrance (it is all lost at faucet)

30 psi x 2.31 ft./psi = 69.3 ft. Head Loss

Square elbows (2) (1 in.)	10 ft.
Open globe valve (1) (1 in.)	25 ft.
Actual pipe length	+ 25 ft.
	60 ft.

$$s = \frac{HL}{L} = \frac{69.3}{60} = 1.155$$

$$Q = .435 \ C \ d^{2.63} \ s^{.54}$$

$$Q = .435 \ 140 \ .0833^{2.63} \ 1.155^{.54}$$

$$Q = .095 \ cfs$$

---

11.   **Answer**   1.9 psi

Draw Diagram. Find equivalent pipe length for fittings;
add to actual pipe length; solve for HL; change to psi.

Swing check (12 in.)	80 ft.
Globe valve (12 in.)	325 ft.
Square els (2) (12 in.)	136 ft.
Equiv. length for fittings	541 ft.

541 ft. + 1000 ft. actual pipe length = 1541 ft. total pipe length.

10 yr. old CIP: C=110

$$Q = .435 \ C \ d^{2.63} \ s^{.54}$$

$$2 = .435 \ 110 \ 1 \ s^{.54}$$

$$.0028 = s$$

HL = .0028 ft./ft. x 1541 ft. = 4.3 ft.

$$\frac{4.3 \ ft.}{2.31 \ ft./psi} = 1.9 \ psi$$

12. **Answer**   A. 9.45 ft.
                 B. 9.43 ft.

Draw Diagram.

A. Solve for Velocity Head. Solve for flow, then HL
   using actual length of pipe. Equate Vh to entrance
   HL; add to HL in straight pipe.

$$Vh = \frac{V^2}{2g} = \frac{5^2}{64.4} = .388 \text{ ft.}$$

$$Q = A \quad V$$

$$Q = .785 \times .83^2 \times 5$$

$$Q = 2.7 \text{ cfs}$$

$$Q = .435 \ C \ d^{2.63} \ s^{.54}$$

$$2.7 = .435 \ 120 \ .83^{2.63} \ s^{.54}$$

$$.0103 = s$$

HL = .0103 ft./ft. x 900 ft. = 9.253 ft.

Head loss for Ordinary entrance = ½ Vh

HL for this fixture = .194 ft.

HL (straight pipe) + HL (fixture) = total HL

9.253 + .194 = 9.45 ft.

B. Obtain equivalent length straight pipe for ordinary entrance; add to pipe length;
   solve for flow, HL.

Straight pipe (10 in.)	900 ft.
Ordinary entrance (10 in.)	+ 15 ft.
	915 ft. total pipe length

Flow with this fitting, from above = 2.7 cfs.

Slope, from above = .0103 ft./ft.

HL = .0103 ft./ft. x 915 ft. = 9.43 ft.

13. **Answer**   1.4%

Obtain equivalent length for fittings; add to pipe length for total; divide this into
equivalent length for fittings.

Pipe length: 2 miles x 5280 ft./mile = 10,560 ft.

45 degree elbows (3) (24 in.)   90 ft.

Gate valves, open (4) (24 in.)   + 60 ft.

150 ft. equiv. length for fittings

150 ft. (fittings) + 10,560 ft. (pipe) = 10,710 ft. total

$\dfrac{150}{10710}$ = .014 = 1.4%

Equivalent length for fittings is 1.4% of total length. Therefore, head loss due to fittings is 1.4% of total head loss. It would be considered negligible.

14. **Answer**   .62 cfs

Using Bernoulli's formula, solve for total HL.   Obtain pipe length; obtain equivalent pipe length for fittings; add; solve for slope, then flow.

Top gage: 100 psi x 2.31 ft./psi = 231 ft.

Bottom gage: 106.5 psi x 2.31 ft./psi = 246 ft.

Single diameter pipe: Velocity Heads cancel.

Ph + Z = Ph + Z + HL

231 + 40 = 246 + 0 + HL

25 ft. = HL

Six runs of 47 ft. length + 40 ft. vertical length = 322 ft. pipe.

Standard elbows (3) (4 in.)   33 ft.

Close return bends (5) (4 in.)   125 ft.

½ closed gate valve (4 in.)   65 ft.

Straight pipe   + 322 ft.

545 ft. total length

New copper pipe = C = 140

$s = \dfrac{HL}{L} = \dfrac{25}{545}$ = .0459 ft./ft.

$Q = .435 \ C \ d^{2.63} \ s^{.54}$

$Q = .435 \ \ 140 \ \ .33^{2.63} \ \ .0459^{.54}$

Q = .62 cfs

15. **Answer**   795 ft.

Tank #2 surface elevation will be Tank #1 surface elevation, minus the HL; add up pipe lengths; convert fittings to equivalent pipe; add on. Change flow to cfs; solve for HL; subtract from Tank #1 surface elevation.

Straight pipe (20 ft. + 50 ft. + 100 ft. + 500 ft. + 1000 ft.)	1670 ft.
Gate valves (3) (12 in.)	21 ft.
Ordinary entrance (12 in.)	18 ft.
Square elbows (3) (12 in.)	204 ft.
Tee (12 in.)	70 ft.
Check valves (2) (12 in.)	160 ft.
45 degree elbow (12 in.)	15 ft.
Ordinary exit (12 in.)	+ 30 ft.
	2188 ft. total pipe length

$$\frac{800 \text{ gpm}}{60 \times 7.48} = 1.78 \text{ cfs}$$

$$Q = .435 \ C \ d^{2.63} \ s^{.54}$$

$$1.78 = .435 \ 110 \ 1 \ s^{.54}$$

$$.0023 = s$$

HL = .0023 ft./ft. x 2188 ft. = 5 ft.

Surface elev. Tank #1	800 ft.
Head loss	- 5 ft.
	795 ft. Surface elev. Tank #2

16. **Answer**   8.9 psi

Draw Diagram. Obtain equivalent length for fittings; add to pipe length; solve for head loss; include elevation head.

10 year old CIP (C=110)

Borda entrance (6 in.)	18 ft.
Std. elbow (1) (6 in.)	17 ft.
Foot valve (swing check) (6 in.)	40 ft.
Actual pipe length	+ 20 ft.
	95 ft.

Flow = 200 gpm x 1440 min. day = 288,000 gpd

.288 MGD x 1.55 cfs/MGD = .45 cfs

$$Q = .435 \ C \ d^{2.63} \ s^{.54}$$

$$.45 = .435 \ 110 \ .5^{2.63} \ s^{.54}$$

$$.0052 = s$$

.0052 ft./ft. x 95 ft. = .5 ft. head loss (from friction and fittings)

Including 20 ft. of elevation head loss in the lift, total head loss = 20.5 ft.

$$\frac{20.5 \text{ ft.}}{2.31 \text{ ft./psi}} = 8.9 \text{ psi pressure drop}$$

17. **Answer**   18.1 ft.

Add pipe lengths; convert fittings to straight pipe length; add on; solve for HL.

Straight pipe (25 ft. + 10 ft. + 65 ft. +		
10 ft. + 20 ft. + 10 ft. + 10 ft.)	150	ft.
Ordinary entrance (8 in.)	12.5	ft.
Gate valve (8 in.)	4.5	ft.
Standard elbows (6) (8 in.)	132	ft.
Ordinary Exit (8 in.)	+ 25	ft.
	324	ft. total pipe L

$$Q = .435 \ C \ d^{2.63} \ s^{.54}$$

$$3.2 = .435 \ 100 \ .67^{2.63} \ s^{.54}$$

$$.056 = s$$

$$HL = .056 \ ft./ft. \times 324 \ ft. = 18.1 \ ft.$$

---

18. **Answer**   .85 cfs

Obtain equivalent length for fittings; add to pipe length; solve for feet of head loss through top section; will be the same for the loop (parallel pipes); solve for flow through loop.

Run of std. tee (1)	12 ft.
Std. tee through side (2) (6 in.)	70 ft.
Std. elbows (2) (6 in.)	34 ft.
Open gate (1) (6 in.)	4 ft.
Half closed gate (1) (6 in.)	100 ft.
Actual pipe length	+ 3000 ft.
	3220 ft.

Head loss through top = 80 psi - 60 psi = 20 psi

20 psi x 2.31 ft./psi = 46.2 ft. head loss

Therefore, head loss through bottom = 46.2 ft.

$$s = \frac{HL}{L} = \frac{46.2}{3220} = .0144 \ ft./ft.$$

$$Q = .435 \ C \ d^{2.63} \ s^{.54}$$

$$Q = .435 \ 120 \ .5^{2.63} \ .0144^{.54}$$

$$Q = .85 \ cfs$$

19. **Answer**   80.2 psi

Draw Diagram. Convert fittings to equivalent pipe;
add on. Change flow to cfs; solve for head loss; add
to lift to obtain pumping head.

$$\frac{400 \text{ gpm}}{7.48 \text{ gal./cu.ft.} \times 60 \text{ sec./min.}} = .891 \text{ cfs}$$

Suction Side

pipe length	10 ft.
90° elbow (6 in.)	34 ft.
check valve (6 in.)	40 ft.
Borda entrance (6 in.)	+ 16 ft.
	100 ft.

$Q = .435 \ C \ d^{2.63} \ s^{.54}$

$.891 = .435 \ 90 \ .5^{2.63} \ s^{.54}$

$.0265 = s$

.0265 ft./ft. x 100 ft. = 2.7 ft. HL

Discharge Side

pipe length	600 ft.
90° elbow (6 in.)	17 ft.
gate valve (6 in.)	3.5 ft.
ordinary exit (6 in.)	+ 16 ft.
	636.5 ft.

$Q = .435 \ C \ d^{2.63} \ s^{.54}$

$.891 = .435 \ 140 \ .5^{2.63} \ s^{.54}$

$.0117 = s$

.0117 ft./ft. x 636.5 ft. = 7.5 ft. HL

175.0 ft. lift	
2.7 ft. suction head loss	
+ 7.5 ft. discharge head loss	
185.2 ft. total pumping head	

$$\frac{185.2 \text{ ft. pumping head}}{2.31 \text{ ft./psi}} = 80.2 \text{ psi}$$

20. **Answer**   A. 29.6 ft.
                  B. 2.55 ft./sec.

Draw Diagram.

A. Solve for HL in each pipe; add; only fitting is
sudden enlargement (about 13 ft. of equiv. pipe
length -ignore it.)

$$Q = .435 \ C \ d^{2.63} \ s^{.54}$$

$$2 = .435 \ 100 \ .67^{2.63} \ s^{.54}$$

$$.0235 = s$$

$$HL = .0235 \ ft./ft. \times 1200 \ ft. = 28.2 \ ft.$$

$$Q = .435 \ C \ d^{2.63} \ s^{.54}$$

$$2 = .435 \ 140 \ 1 \ s^{.54}$$

$$.0018 = s$$

$$HL = .0018 \ ft./ft. \times 800 \ ft. = 1.4 \ ft.$$

Adding HL's: 28.2 ft. + 1.4 ft. = 29.6 ft. total HL

B. $Q = A \ V$

$$2 = .785 \times 1 \ V$$

$$2.55 \ ft./sec. = V$$

---

21. **Answer**   2.1 cfs

Draw Diagram. Convert fittings to lengths of
equivalent pipe; add to connecting straight pipe;
solve for length of a pipe equivalent to the
system (assume Q, solve for HL in each pipe,
add HL's; create equiv. pipe, solve for slope,
solve for length). Using given HL and L of equiv., solve for slope, then flow.

8 inch pipe	
Straight pipe	1000 ft.
Ordinary entrance	12 ft.
Gate Valve	5 ft.
Elbows (3)	66 ft.
Globe valve	+ 200 ft.
	1303 ft.

6 inch pipe

Straight pipe	50 ft.
Sudden contraction	5 ft.
Angle valve	90 ft.
Tee	12 ft.
Ordinary exit	+ 18 ft.
	175 ft.

Assume Q = 3 cfs.

8 inch

$$Q = .435 \ C \ d^{2.63} \ s^{.54}$$

$$3 = .435 \ \ 100 \ \ .67^{2.63} \ s^{.54}$$

$$.0497 = s$$

HL = .0497 ft./ft. x 1303 ft. = 64.8 ft.

6 inch

$$Q = .435 \ C \ d^{2.63} \ s^{.54}$$

$$3 = .435 \ \ 100 \ \ .5^{2.63} \ s^{.54}$$

$$.2068 = s$$

HL = .2068 ft./ft. x 175 ft. = 36.2 ft.

Adding HL's: 64.8 ft. + 36.2 ft. = 101 ft. total HL (from assumed flow)

Choose equivalent pipe: 6 inch diameter C=100.

$$s = \frac{HL}{L}$$

$$.2068 = \frac{101}{L}$$

$$488.4 \ ft. = L$$

Use L of equivalent pipe and given HL - solve for slope.

$$s = \frac{HL}{L} = \frac{50}{488.4} = .1024 \ ft./ft.$$

$$Q = .435 \ C \ d^{2.63} \ s^{.54}$$

$$Q = .435 \ \ 100 \ \ .5^{2.63} \ s^{.54}$$

$$Q = 2.1 \ cfs$$

chapter 9

1. **Answer**   .0031 ft./ft.

   Draw Diagram. Solve for R, then slope.

   $$R = \frac{WA}{WP} = \frac{6.25 \times 8}{6.25 + 8 + 6.25} = 2.44 \text{ ft.}$$

   $$V = \frac{1.486}{n} \ R^{.66} \ s^{.5}$$

   $$14.9 = \frac{1.486}{.01} \ 2.44^{.66} \ s^{.5}$$

   .0031 ft./ft. = s

2. **Answer**   .4 ft.

   Draw Diagram. Solve for R, then Vel., then slope, then drop.

   $$R = \frac{WA}{WP} = \frac{7 \times 15}{7 + 15 + 7} = 3.62 \text{ ft.}$$

   $Q = A \ V$

   $600 = 15 \times 7 \times V$

   $5.71 \text{ ft./sec.} = V$

   $$V = \frac{1.486}{n} \ R^{.66} \ s^{.5}$$

   $$5.71 = \frac{1.486}{.012} \ 3.62^{.66} \ s^{.5}$$

   .0004 ft./ft. = s

   .0004 ft./ft. x 1000 ft. = .4 ft. of incline

3. **Answer**   .00009 ft./ft.

   Draw Diagram. Solve for area, then width and depth; solve for R, then slope.

   $Q = A \ V$

   $900 = A \times 3$

   300 sq.ft. = A

Area $=$ width x depth

$300 = 3X \times X$

10 ft. $= X$

$R = \dfrac{WA}{WP} = \dfrac{10 \times 30}{10 + 30 + 10} = 6$ ft.

$V = \dfrac{1.486}{n} \ R^{.66} \ s^{.5}$

$3 = \dfrac{1.486}{.015} \ 6^{.66} \ s^{.5}$

$.00009$ ft./ft. $= s$

---

4. **Answer**   A. 2.3 cfs
                  B. 2.6 cfs

A. <u>Manning</u>: Solve for R, then Vel., then flow.

$R = \dfrac{d}{4} = \dfrac{1}{4} = .25$ ft.

$V = \dfrac{1.486}{n} \ R^{.66} \ s^{.5}$

$V = \dfrac{1.486}{.013} \ .25^{.66} \ .004^{.5}$

$V = 2.89$ ft./sec.

$Q = A \ \ V$

$Q = .785 \times 1 \times 2.89$

$Q = 2.3$ cfs

B. <u>Hazen-Williams</u>: C factor CIP 130 → 100. Use 120; solve for flow.

$C = 120$

$Q = .435 \ \ C \ \ d^{2.63} \ \ s^{.54}$

$Q = .435 \ \ 120 \ \ 1 \ \ .004^{.5}$

$Q = 2.6$ cfs

5. **Answer**   692 cfs

Draw Diagram. Assume unrestricted flow from the
reservoir. HL is equivalent to total slope of channel
plus extra depth at reservoir. Solve for slope, then R,
then velocity, then flow.

$$HL = .72 \text{ ft. drop of channel} + 6.22$$
$$\text{ft.} - 4.12 \text{ ft. depth of reservoir} = 2.82 \text{ ft.}$$

$$s = \frac{HL}{L} = \frac{2.82}{800} = .0035 \text{ ft./ft.}$$

$$R = \frac{WA}{WP} = \frac{4.12 \times 15}{4.12 + 15 + 4.12} = 2.66 \text{ ft.}$$

$$V = \frac{1.486}{n} \quad R^{.66} \quad s^{.5}$$

$$V = \frac{1.486}{.015} \quad 2.66^{.66} \quad .0035^{.5}$$

$$V = 11.2 \text{ ft./sec.}$$

$$Q = A \quad V$$

$$Q = 4.12 \times 15 \times 11.2$$

$$Q = 692 \text{ cfs}$$

6. **Answer**   63.9 cfs

Draw Diagram. Solve for R, then Vel., then flow.

$$R = \frac{d}{4} = \frac{4}{4} = 1 \text{ ft.}$$

$$V = \frac{1.486}{n} \quad R^{.66} \quad s^{.5}$$

$$V = \frac{1.486}{.016} \quad 1 \quad .003^{.5}$$

$$V = 5.09 \text{ ft./sec.}$$

$$Q = A \quad V$$

$$Q = .785 \times 4^2 \times 5.09$$

$$Q = 63.9 \text{ cfs}$$

7. **Answer**   118,483 gal./4 hrs.

Draw Diagram.

$$Q = A \quad V$$

$$Q = .83 \times .67 \times 2$$

$$Q = 1.1 \text{ cfs}$$

1.1 cuft/sec. x 60 sec./min x 60 min./hr. x
7.48 gal./cuft x 4 hr. = 118,483 gal./4 hr.

8. **Answer**   6.5 ft.

Draw Diagram.   Solve for V, then Q at
s=.0144.   Create a comparison chart   at
different depths; solve for Vel. through both
Mannings and Q=AV. At a given depth, if the
Vel. comes out the same through both formulas,
then that must be the right depth.

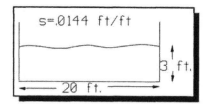

$$s = .0144$$

$$R = \frac{WA}{WP} = \frac{3 \times 20}{3 + 20 + 3} = 2.31 \text{ ft.}$$

This is R at a depth of 3 ft. - which the slope of .0144 ft./ft. creates.

$$V = \frac{1.486}{n} \quad R^{.66} \quad s^{.5}$$

$$V = \frac{1.486}{.013} \quad 2.31^{.66} \quad .0144^{.5}$$

$$V = 23.8 \text{ ft./sec.}$$

$$Q = A \quad V$$

$$Q = 20 \times 3 \times 23.8$$

$$Q = 1428 \text{ cfs}$$

$$s = .00144$$

Depth Trials:
Assume a depth: using it, solve for Velocity using Mannings (s=.00144),
then using Q=AV. When velocities match, then depth is correct depth.

Depth (ft.)	Velocity (ft./sec.) (Mannings)	Velocity (ft./sec.) (Q=AV)
10	12.5	7.1
15	14.1	4.8
6.5	10.7	10.9

9.   **Answer**   .0004 ft./ft.

Draw Diagram.   There are perhaps three different
ways to handle this; answers should be similar.

a.   Since it is an open channel flow problem, use
     Mannings formula; choose .013 as an average n factor,
     equatable to C-100.

$$V = \frac{1.486}{n}\ R^{.66}\ s^{.5}$$

$$4.2 = \frac{1.486}{.013}\ \left(\frac{3.9 \times 12.5}{3.9 + 12.5 + 3.9}\right)^{.66}\ s^{.5}$$

.0004 ft./ft. = s

b.   Choose a different version of Hazen Williams formula, one which employes hydraulic
     radius instead of diameter.

c.   Use our version of Hazen Williams formula; calculate for diameter by obtaining flow
     and inserting into Q = A V as if it were a full pipe. (This method has the least validity,
     for the hydraulics of rectangular open channels is not the same as full pipes).

10.   **Answer**   Trapezoid carries greater flow.

Both are laid on the same slope - so assume any slope. Solve for R, then Velocity,
then flow in each.

Assume slope = 1

Rectangle:
$$R = \frac{WA}{WP} = \frac{9 \times 20}{9 + 20 + 9} = 4.74 \text{ ft.}$$

$$V = \frac{1.486}{n}\ R^{.66}\ s^{.5}$$

$$V = \frac{1.486}{.015}\ 4.74^{.66}\ 1$$

V = 277 ft./sec.

Q = A   V

Q = 9 x 20 x 277

Q = 49860 cfs

Trapezoid:
$$R = \frac{wh + yh^2}{w + 2h\sqrt{y^2 + 1}} = \frac{(20 \times 6) + (1.33 \times 6^2)}{20 + (2 \times 6)\sqrt{1.33^2 + 1}} = 3.15 \text{ ft.}$$

$$V = \frac{1.486}{n} \quad R^{.66} \quad s^{.5}$$

$$V = \frac{1.486}{.01} \quad 3.15^{.66} \quad 1$$

$$V = 317 \text{ ft./sec.}$$

$$Q = A \quad V$$

$$Q = wh + yh^2 \quad V$$

$$Q = (20 \times 6) + (1.33 \times 6^2) \times 317$$

$$Q = 53218 \text{ cfs}$$

---

11. **Answer**   A. .2283 ft.
                  B. .341 sq.ft.

Draw Diagram.

A. Obtain % fullness; refer to Hydraulic Elements
    Curve for % R of full pipe; solve for R of full
    pipe; multiply by % value for our pipe.

   % fullness = 6/10 = 60%

   From curve, R for a 60% full pipe is 110% of R of a full pipe.

   R (full pipe) $= \frac{d}{4} = \frac{.83}{4} = .2075$ ft.

   .2075 ft. (R-full) x 110% (% of full value) = .2283 ft. (R, semi-full pipe)

B. Do the same thing for Area.

   Area of 60% full pipe = 63% area of full pipe.

   A (full pipe) = .785 d^2

   A = .785 x .83^2

   A = .54 sq.ft.

   .54 sq.ft. (A - full) x .63 (% of full value) = .341 sq.ft. (A, semi-full pipe)

---

12. **Answer**   2 ft./sec.

Change gpm to cfs; refer to Hyd.El.Curve for % A; obtain value for A; solve for
Vel.

   $\frac{135 \text{ gpm}}{60 \times 7.48} = .3$ cfs

   33% full pipe: A = 28% of A (full pipe) - from curve.

$$A \text{ (full)} = .785 \quad d^2$$

$$A = .785 \times .83^2$$

$$A = .54 \text{ sq.ft.}$$

.54 sq.ft. (A, full) x .28 (% of full value) = .15 sq.ft. (A, semi-full pipe)

$$Q = A \quad V$$

$$.3 = .15 \times V$$

$$2 \text{ ft./sec.} = V$$

---

13. **Answer**   7.2 inches deep

Change flow to cu.ft./sec.; solve for discharge of full 24 inch diameter pipe. Consult hydraulic elements curve; come up from that % to discharge curve, then left to depth of flow; calculate water depth.

$$\frac{550 \text{ gal./min.}}{7.48 \text{ gal./cu.ft.} \times 60 \text{ sec./min.}} = 1.23 \text{ cfs (semi-full pipe)}$$

$$Q = A \quad V$$

$$Q = .785 \times 2^2 \quad 2$$

$$Q = 6.28 \text{ cfs (full pipe)}$$

$$\frac{1.23 \text{ cfs}}{6.28 \text{ cfs}} = .196 = 19.6\% \text{ of full pipe flow}$$

From curve, a pipe carrying 19.6% of full pipe flow is 30% full.

24 inch diameter x .3 = 7.2 inches deep

---

14. **Answer**   3.1 ft./sec.

Draw Diagram.   Calculate slope; with Hyd. El. Curve, calculate R; use Mannings formula to solve for velocity.

$$\text{Slope} = \frac{5 \text{ ft. incline}}{1500 \text{ ft. length}} = .0033 \text{ ft./ft.}$$

Concrete pipe n = .015

$$\% \text{ Depth of flow} = \frac{10 \text{ in.}}{18 \text{ in.}} = .56 = 56\%$$

At 56% fullness, R = 106% of full pipe R

$$R \text{ (full pipe)} = \frac{d}{4} = \frac{1.5}{4} = .375 \text{ ft.}$$

$$R \text{ (Semi-full)} = 1.06 \times .375 \text{ ft.}$$

$$R \text{ (Semi-full)} = .398 \text{ ft.}$$

$$V = \frac{1.486}{n} R^{.66} s^{.5}$$

$$V = \frac{1.486}{.015} .398^{.66} .0033^{.5}$$

$$V = 3.1 \text{ ft./sec.}$$

---

15. **Answer**   2.2 ft./sec.

Inspecting Hydraulic Elements Curve, velocity and hydraulic radius of a half full pipe are the same as that of a full pipe. Use Mannings formula; treat the pipe as full.

$$V = \frac{1.486}{n} R^{.66} s^{.5}$$

$$V = \frac{1.486}{.025} \left(\frac{2}{4}\right)^{.66} \left(\frac{1}{800}\right)^{.5}$$

$$V = 2.2 \text{ ft./sec.}$$

---

16. **Answer**   A. 16 inch diameter
                   B. 57.3 ft.

A. A pipe size is needed that will hold the flow at 80% fullness (or lss). Assume a desired velocity; solve for area of flow with this velocity and given flow; consult Hydraulic Element Curve for % of full pipe area; solve for full pipe area and diameter.

$$2 \text{ MGD} \times 1.55 \text{ cfs/MGD} = 3.1 \text{ cfs}$$

Scouring velocity is needed; use 3 ft./sec.

$$Q = A \quad V$$

$$3.1 = A \quad 3$$

$$1.03 \text{ sq.ft.} = A \quad \text{This is true area of flow}$$

From Hydraulic Elements Curve, at 80% depth of flow, the area is about 82% of area of a full pipe.

$$1.03 \text{ sq.ft. (semi-full area)} = .82 A \text{ (full pipe area)}$$

$$1.25 \text{ sq.ft.} = A \text{ (full pipe area)}$$

A (full pipe) = .785 d²

1.25 = .785 d²

1.26 ft. = d

16 in. = d

Use a 16 in. diameter pipe and it will have about 80% depth of flow at 2 MGD

B. Calculate R; solve for slope, then incline.

From Hydraulic Elements Curve, R for an 80% full pipe is 123% of full pipe R.

$$R \text{ (full pipe)} = \frac{d}{4} = \frac{1.3}{4} = .325 \text{ ft.}$$

R (full pipe) x 1.23 = R (semi-full)

.325 x 1.23 = R (semi-full)

.4 = R (semi-full)

Concrete pipe n = .015

$$V = \frac{1.486}{n} \quad R^{.66} \quad s^{.5}$$

$$V = \frac{1.486}{.015} \quad .4^{.66} \quad s^{.5}$$

.0031 = s

.0031 ft./ft. x 5280 ft./mi. x 3.5 mi. = 57.3 ft. incline over the 3.5 miles

17. **Answer**   9.3 ft./sec.

Draw Diagram.   Solve for R, then V.

$$R = \frac{yh}{2\sqrt{y^2 + 1}}$$

$$R = \frac{4 \times 6}{2\sqrt{4^2 + 1}}$$

R = 2.91 ft.

$$V = \frac{1.486}{n} \quad R^{.66} \quad s^{.5}$$

$$V = \frac{1.486}{.025} \quad 2.91^{.66} \quad .006^{.5}$$

V = 9.3 ft./sec.

18. **Answer**   .31 cfs

Draw Diagram. Solve for slope; using Hyd.El.Curve, solve for R, then Velocity. Using curve again, solve for Area, then flow.

$$\% \text{ fullness} = \frac{3}{10} = 30\% \text{ full}$$

With curve, 30% full pipe: R = 67% of R for full pipe.

$$R \text{ (full pipe)} = \frac{d}{4} = \frac{.83}{4} = .2075 \text{ ft.}$$

.2075 ft. (R, full) x .67 (% of full value) = .139 ft. (R, semi-full pipe)

$$V = \frac{1.486}{n} \quad R^{.66} \quad s^{.5}$$

$$V = \frac{1.486}{.025} \quad .139^{.66} \quad .02^{.5}$$

$$V = 2.29 \text{ ft./sec.}$$

With curve, 30% full pipe: A = 25% of A for full pipe.

$$A \text{ (full)} = .785 \quad d^2$$

$$A = .785 \times .83^2$$

$$A = .54 \text{ sq.ft.}$$

.54 sq.ft. (A, full) x .25 (% of full value) = .135 sq.ft. (A, semi-full pipe)

$$Q = A \quad V$$

$$Q = .135 \times 2.29$$

$$Q = .31 \text{ cfs}$$

---

19. **Answer**   32 inches deep

Draw Diagram. Solve for slope. There is too little data to solve for depth directly either by Q=AV or by Mannings formula. If the correct depth of flow was known, velocity could be obtained by either formula, and it should be the same velocity using either one. Assume different depths; create a chart. Compare velocities which each formula yields. When they are the same, then that is the correct depth.

Depth	% Fullness	R	V (Manning)	V (Q=AV)	A
24 in.	50%	1	4.5	6.9	6.28
36 in.	75%	1.2	5.1	4.3	10
48 in.	100%	1	4.5	3.4	12.6
32 in.	65%	1.1	4.9	4.9	8.2

It is necessary to work with the Hyd.El.Curve through % fullness to a true value for R & A for the assumed depth.

After trying 50%, 75%, 100% fullness, and comparing velocities, 65% was attempted.

---

20. **Answer**   A. 14 inch diam.
                  B. Max. Q: 10.4 inches deep; Av. Q: 6 inches deep
                  C. Max. Q: Vel. = 3 ft./sec.; Av. Q: Vel. = 2.3 ft./sec.

A. Sanitary sewer must have scouring velocity. Start with Vel. = 3 ft./sec. Since max. flow is 2.5 cfs, let pipe be full at this flow. Solve for diameter.

$$Q = A \quad V$$

$$2.5 = .785 \quad d^2 \quad 3$$

$$1.03 \text{ ft.} = d$$

Choose 14 inch pipe. (1.167 ft.)

B. Pipe must be sloped enough to keep the velocity at 3 ft./sec. The area of flow is based on the full pipe diameter calculated above; solve for this area of flow. Get its % of the full area of the pipe; refer to Hyd. Elements Curve for % depth of flow.

$$A = .785 \quad d^2$$

$$A = .785 \times 1.03^2$$

$$A = .833 \text{ sq.ft.}$$

$$A = .785 \quad d^2$$

$$A = .785 \times 1.167^2$$

$$A = 1.07 \text{ sq.ft.}$$

$$\frac{.833 \text{ sq.ft.}}{1.07 \text{ sq.ft.}} = .78 = 78\% \text{ of area of pipe is submerged}$$

From Hyd. Elements Curve 78% of area = 74% depth of flow.

.74 x 1.167 ft. diam. = .864 ft. depth = 10.4 inches deep

Average daily flow = 1cfs. This is 40% of max. flow (2.5 cfs).

Max. flow fills 74% depth of pipe.

From Hyd. Elements Curve, Max. flow, at this depth, is 90% of the flow of this pipe full.

Therefore, the average flow = 40% x 90% of flow of full pipe = 36% of flow of full pipe.

Again, from curve, at 36% of flow of full pipe, (our average flow), the pipe holds 43% depth of flow.

.43 x 1.167 ft. diam. .5 ft. depth = 6 inches deep

C. Max flow: 78% of area of full pipe area. Solve for velocity.

$Q = A \quad V$

$2.5 = .78 \times .785 \times 1.167^2 \quad V$

$3 \text{ ft./sec.} = V$

Note: Area of flow is the same as in original pipe diameter calculation; therefore, velocity is the same.

Average flow: at 43% depth of flow in pipe, area (from curve) is 41% of area of full pipe. Solve for velocity.

$Q = A \quad V$

$1 = .41 \times .785 \times 1.167^2 \quad V$

$2.3 \text{ ft./sec.} = V$

21. **Answer**   2.48 ft.

Draw Diagram. Since channel bottom is horizontal, water surface will create a slope which is equivalent to the HL. Depth upstream = HL/1000 ft. plus the 2 ft. obstruction.

$R = \dfrac{WA}{WP} = \dfrac{6 \times 2}{2 + 6 + 2} = 1.2 \text{ ft.}$

10 MGD = 15.5 cfs

$Q = A \quad V$

$15.5 = 2 \times 6 \quad V$

$1.3 \text{ ft./sec.} = V$

$$V = \frac{1.486}{n} \quad R^{.66} \quad s^{.5}$$

$$1.3 = \frac{1.486}{.02} \quad 1.2^{.66} \quad s^{.5}$$

$$.00024 \text{ ft./ft.} = s$$

$$HL = .00024 \text{ ft./ft. x } 2000 \text{ ft.} = .48 \text{ ft.}$$

$$.48 \text{ ft. HL} + 2 \text{ ft. for obstruction} = 2.48 \text{ ft.}$$

---

22. **Answer**   4 ft.

Solve for slope; solve for R, then diameter.

$$s = \frac{HL}{L} = \frac{3}{1500} = .002 \text{ ft./ft.}$$

$$V = \frac{1.486}{n} \quad R^{.66} \quad s^{.5}$$

$$5 = \frac{1.486}{.013} \quad R^{.66} \quad .002^{.5}$$

$$1 \text{ ft.} = R$$

$$R = \frac{d}{4}$$

$$1 = \frac{d}{4}$$

$$4 \text{ ft.} = d$$

---

23. **Answer**   4039 gpm

Draw Diagram. HL = 3 ft. of slope plus 5 ft. of pressure in sump = 8 ft. HL. Solve for slope, then R, then velocity, then flow; change to gpm.

$$S = \frac{HL}{L} = \frac{8}{150} = .0533 \text{ ft./ft.}$$

$$R = \frac{d}{4} = \frac{1}{4} = .25 \text{ ft.}$$

$$V = \frac{1.486}{n} \quad R^{.66} \quad s^{.5}$$

$$V = \frac{1.486}{.012} \quad .25^{.66} \quad .0533^{.5}$$

$$V = 11.5 \text{ ft./sec.}$$

$$Q = A \quad V$$
$$Q = .785 \times 1 \times 11.5$$
$$Q = 9 \text{ cfs}$$

9 cfs x 60 sec./min. x 7.48 gal./cu.ft. = 4039 gpm

24. **Answer**   A. 12 ft. diameter
                    B. 8.4 ft. depth

A. Solve for area of flow; choose diameter.

$$Q = A \quad V$$
$$125 = A \times 1.5$$
$$83.3 \text{ sq.ft.} = A \qquad \text{This is area of \underline{flow}.}$$

$$A = .785 \quad d^2$$
$$83.3 = .785 \quad d^2$$
$$10.3 \text{ ft.} = d \qquad \text{This is the minimum pipe size that would hold this flow.}$$

Therefore choose 12 ft. diameter pipe.

B. Solve for Area of chosen pipe; solve for % of area that our flow uses in this pipe; consult Hyd. El.Curve for % fullness; solve for depth.

$$A = .785 \quad d^2$$
$$A = .785 \times 12^2$$
$$A = 113 \text{ sq.ft.}$$

$$\frac{83.3 \text{ sq.ft.}}{113 \text{ sq.ft.}} = .74 = 74\% \text{ Semi-full flow is 74\% of full pipe area.}$$

From curve, pipe = 70% full.

12 ft. diam. pipe x .7 = 8.4 ft. depth of flow

Check velocity with Manning's formula: 3.2 ft./sec.

25. **Answer**   Variable

Assumptions: need to have these
Pipe material choice: cement lined steel pipe (n=.013)
100 MGD = 155 cfs = pipe flowing full
Wish to have scouring velocity: Vel. = 3 ft./sec.
Solve for diameter:

$$Q = A \quad V$$

$$155 = .785 \times d^2 \times 3$$

$$8.113 \text{ ft.} = d$$

Choose 8.5 ft. diameter pipe

Calculate slope for pipe construction, using our chosen diameter pipe; obtain slope that will produce 3 ft./sec. velocity; even though this is no longer a full pipe, the area of flow is the same, therefore the velocity remains the same. However, R is not that of a full pipe; compare areas to obtain depth of flow. Get % R compared to that of a full pipe. Obtain R for semi-full pipe; then solve for slope.

Area of flow = $.785 \times 8.113^2$ = 51.67 sq.ft.

Area of our pipe = $.785 \times 8.5^2$ = 56.72 sq.ft.

$\dfrac{51.67}{56.72}$ = .91 = 91% of full pipe area

Referring to Hyd. Elements Curve, with 91% of full pipe area, pipe is 86% full.

From curve, a pipe 86% full has an R that is 122% of full pipe R.

Full pipe R = $\dfrac{d}{4}$ = $\dfrac{8.5}{4}$ = 2.125 ft.

Semi-full pipe R = 1.22 x 2.125 ft. = 2.6 ft.

Vel. = $\dfrac{1.486}{.013}$ $R^{.66}$ $s^{.5}$

$3 = 114.3 \quad 2.6^{.66} \quad s^{.5}$

.0002 ft./ft. = s

.0002 ft./ft. x 5280 ft./mile x 150 miles = 158.4 ft. drop in elevation needed for pipeline.

chapter 10

1. **Answer**   21.6 ft. of water

Sharp edged orifice C_d = .6; solve for flow; solve for head.

$$Q = A \quad V$$

$$Q = .785 \times .167^2 \times 22.5$$

$$Q = .49 \text{ cfs}$$

$$Q = C_d \quad A \quad \sqrt{(Ph_1 - Ph_2) \times 2g}$$

$$.49 = .6 \times .785 \times .167^2 \quad \sqrt{(Ph_1 - Ph_2) \times 64.4}$$

$$21.6 \text{ ft.} = Ph_1 - Ph_2$$

2. **Answer**   73.3 ft./sec.

Short tube $C_d = .8$; change pressure to feet of water; solve for flow; then velocity.

$$60 \text{ psi} \times 2.31 \text{ ft./psi} = 138.6 \text{ ft.}$$

$$Q = C_d \quad A \quad \sqrt{(Ph_1 - Ph_2) \times 2g}$$

$$Q = .8 \times .785 \times .0417^2 \quad \sqrt{(138.6 - 0) \times 64.4}$$

$$Q = .1 \text{ cfs}$$

$$Q = A \quad V$$

$$.1 = .785 \times .0417^2 \quad V$$

$$73.3 \text{ ft./sec.} = V$$

3. **Answer**   20 ft.

Draw diagram.   Hole in tank is an orifice; $C_d = .8$; solve for depth.

$$.1 \text{ MGD} \times 1.55 \text{ cfs/MGD} = .155 \text{ cfs}$$

$$Q = C_d \quad A \quad \sqrt{(Ph_1 - Ph_2) \times 2g}$$

$$.155 = .8 \times .785 \times .083^2 \quad \sqrt{(Ph_1 - 0) \times 64.4}$$

$$20 \text{ ft.} = Ph_1$$

4. **Answer**   1,203,800 gpd

Change psi to feet of water; assume round edged orifice $C_d = .9$; solve for flow; change to gpd.

$$60 \text{ psi} \times 2.31 \text{ ft./psi} = 138.6 \text{ ft.}$$

$$Q = C_d \quad A \quad \sqrt{(Ph_1 - Ph_2) \times 2g}$$

$$Q = .9 \quad .785 \times .167^2 \quad \sqrt{(138.6 - 0) \times 64.4}$$

$$Q = 1.86 \text{ cfs}$$

$$1.86 \text{ cfs} \times 60 \text{ sec./min.} \times 1440 \text{ min./day} \times 7.48 \text{ gal./cu.ft.} = 1,203,800 \text{ gpd}$$

5. **Answer**   1.17 cfs

Change pressure to feet of water; solve for flow.

$$60 \text{ psi x } 2.31 \text{ ft./psi} = 138.6 \text{ ft.}$$

$$40 \text{ psi x } 2.31 \text{ ft./psi} = 92.4 \text{ ft.}$$

$$Q = C_d \ A \ \sqrt{\frac{(Ph_1 - Ph_2) \times 2g}{1 - \left(\frac{d_2}{d_1}\right)^4}}$$

$$Q = .98 \times .785 \times .167^2 \ \sqrt{\frac{(138.6 - 92.4) \times 64.4}{1 - \left(\frac{.167}{.67}\right)^4}}$$

$$Q = 1.17 \text{ cfs}$$

---

6. **Answer**   1.2 ft.

Total weir height must be at least equal to depth of water plus head on weir. Change flow to cfs; solve for original depth; solve for head on weir; add.

$$\frac{1000 \text{ gpm}}{60 \text{ sec./min. x } 7.48 \text{ gal./cuft}} = 2.23 \text{ cfs}$$

$$Q = A \ V$$

$$2.23 = 2 \times d \times 5$$

$$.22 \text{ ft.} = \text{depth}$$

$$Q = 2.5 \ h^{2.5}$$

$$2.23 = 2.5 \ h^{2.5}$$

$$.96 \text{ ft.} = h$$

.22 ft. original depth + .96 ft. head on weir = 1.2 ft. tall (weir must be)

---

7. **Answer**   4.3 ft.

Solve for head on weir; add to weir height.

$$Q = 3.33 \ L \ h^{1.5}$$

$$100 = 3.33 \times 20 \times h^{1.5}$$

$$1.3 \text{ ft.} = h$$

3 ft. height of weir + 1.3 ft. head on weir = 4.3 ft. depth.

8. **Answer** .72 ft.

Change gpm to cfs; solve for head on weir.

$$\frac{500 \text{ gpm}}{60 \text{ sec./min. x } 7.48 \text{ gal./cu.ft.}} = 1.1 \text{ cfs}$$

$Q = 2.5 \ h^{2.5}$

$1.1 = 2.5 \ h^{2.5}$

$.72 \text{ ft.} = h$

9. **Answer** .47 inches

Assume a flat weir; change MGD to cfs; solve for head on weir.

.5 MGD x 1.55 cfs/MGD = .775 cfs

$Q = 3.33 \ L \ h^{1.5}$

$.775 = 3.33 \text{ x } 30 \text{ x } h^{1.5}$

$.039 \text{ ft.} = h$

$.039 \text{ ft. x } 12 \text{ in./ft.} = .47 \text{ inches.}$

10. **Answer** 2 inches diameter

Assume sharp edged orifice C_d = .6; assume discharge to atmosphere; solve for diameter.

$Q = C_d \ A \ \sqrt{(Ph_1 - Ph_2) \text{ x } 2g}$

$.565 = .6 \text{ x } .785 \ d^2 \ \sqrt{(28.5 - 0) \text{ x } 64.4}$

$.17 \text{ ft.} = d$

$2 \text{ inches} = d$

11. **Answer** 840 cfs

Change inches water reading to ft. of water; solve for flow.

$$\frac{2 \text{ inches water (reading)}}{12 \text{ in./ft.}} = .167 \text{ ft. of water}$$

$$Q = C_d \ A \ \sqrt{\frac{(Ph_1 - Ph_2) \text{ x } 2g}{1 - \left(\frac{d_2}{d_1}\right)^4}}$$

$$Q = .98 \times .785 \times 16^2 \sqrt{\frac{.167 \times 64.4}{1 - \left(\frac{16}{20}\right)^4}}$$

$Q = 840$ cfs

---

12. **Answer** 5.8 inches each side

Square orifice $C_d = .6$; solve for area of orifice, then side dimensions.

$Q = C_d \ A \ \sqrt{(Ph_1 - Ph_2) \times 2g}$

$5 = .6 \ A \ \sqrt{(20 - 0) \times 64.4}$

$.23$ sq.ft. $= A$

$A = \text{side}^2$

$.23 = s^2$

$.482$ ft. $= s$

5.8 in. $= s$

---

13. **Answer** .054 cfs

Change orifice diameter to feet. Change inches of mercury to ft. of water. Solve for flow.

$\dfrac{1 \text{ inch orifice}}{12 \text{ in./ft.}} = .083$ ft.

$\dfrac{3 \text{ inches mercury}}{12 \text{ in./ft.}} = .25$ feet of mercury

$.25$ ft. mercury x 13.6 (SG mercury) $= 3.4$ ft. water

$Q = C_d \ A \ \sqrt{(Ph_1 - Ph_2) \times 2g}$

$Q = .68 \times .785 \times .083^2 \ \sqrt{3.4 \times 64.4}$

$Q = .054$ cfs

---

14. **Answer** 4.5 ft. of water

Change inches of mercury to feet of water.

$\dfrac{4 \text{ inches mercury}}{12 \text{ in./ft.}} = .33$ ft. mercury

$.33$ ft. mercury x 13.6 (SG of mercury) $= 4.5$ feet of water

15. **Answer**   17.5 inches of mercury

Diameter change in this venturi is large; formula for orifice meters can be used; Convert diameter to feet; solve for pressure differential; change to inches of mercury.

$$\frac{4 \text{ inches}}{12 \text{ in./ft.}} = .33 \text{ ft.}$$

$$Q = C_d \quad A \quad \sqrt{(Ph_1 - Ph_2) \times 2g}$$
$$3 = .98 \times .785 \times .33^2 \quad \sqrt{(Ph_1 - Ph_2) \times 64.4}$$
$$19.8 \text{ ft.} = Ph_1 - Ph_2$$

$$\frac{19.8 \text{ ft. of water}}{13.6 \text{ (SG mercury)}} = 1.46 \text{ ft. of mercury}$$

1.46 ft. mercury x 12 in./ft. = 17.5 inches of mercury

---

16. **Answer**   A. .25 ft.
    B. 3.4 ft.
    C. .32 cfs
    D. 14.6 ft./sec.
    E. 1.6 ft./sec.

Diameter difference is large. Formula for orifice may be used.

A. Change inches of mercury to feet of mercury.

$$\frac{3 \text{ inches mercury}}{12 \text{ in./ft.}} = .25 \text{ ft. mercury}$$

B. Change feet of mercury to feet of water.

.25 ft. mercury x 13.6 (SG mercury) = 3.4 ft. of water

C. Change diameters to feet; solve for flow.

6 inches = .5 ft.

2 inches = .167 ft.

$$Q = C_d \quad A \quad \sqrt{(Ph_1 - Ph_2) \times 2g}$$
$$Q = .98 \times .785 \times .167^2 \quad \sqrt{3.4 \times 64.4}$$
$$Q = .32 \text{ cfs}$$

D. Solve for velocity at throat.

$$Q = A \quad V$$
$$.32 = .785 \times .167^2 \quad V$$
$$14.6 \text{ ft./sec.} = V$$

E. Solve for velocity upstream.

$$Q = A \quad V$$

$$.32 = .785 \times .5^2 \quad V$$

$$1.63 \text{ ft./sec.} = V$$

---

17. **Answer**   21.6 cfs

Change upstream depth to feet; solve for flow; calculate % submergence; adjust flow.

$$\frac{20 \text{ inches}}{12 \text{ in./ft.}} = 1.67 \text{ ft.}$$

$$Q = 4 \quad W \quad H_a^{1.52} \quad W^{.026}$$

$$Q = 4 \times 3 \times 1.67^{1.52} \times 3^{.026}$$

$$Q = 27 \text{ cfs}$$

$$\% \text{ submergence} = \frac{H_b}{H_a} = \frac{18}{20} = .9 = 90\% \text{ submergence}$$

From % submergence graph, when throat is 90% submerged, flow is 80% of calculated flow.

$$27 \text{ cfs} \times .8 = 21.6 \text{ cfs}$$

---

18. **Answer**   33 ft.

Assume $C_d = .6$; change inches to feet; change psi to feet; solve for flow; solve for horizontal distance of free fall.

$$1 \text{ inch} = .083 \text{ ft.}$$

$$80 \text{ psi} \times 2.31 \text{ ft./psi} = 184.8 \text{ ft.}$$

$$Q = C_d \quad A \quad \sqrt{(Ph_1 - Ph_2) \times 2g}$$

$$Q = .6 \times .0833^2 \quad \sqrt{(184.8 - 0) \times 64.4}$$

$$Q = .46 \text{ cfs}$$

$$Q = \frac{4 \quad A \quad X}{\sqrt{y}}$$

$$.46 = \frac{4 \times .0833^2 \times X}{\sqrt{4}}$$

$$33 \text{ ft.} = X$$

19. **Answer**   660,000 gpd

Pipe 1/3 full; solve for area (Hyd. Elements Curve); solve for flow; change to gpd.

From Hydraulic Elements Curve, pipe 1/3 full has an area which is 28% of the area of a full pipe.

$$A_{full} = .785 \ d^2$$
$$A_{full} = .785 \times 1^2$$
$$A_{full} = .785 \ sq.ft.$$

$$.28 \times .785 \ sq.ft. = .22 \ sq.ft. \ (area \ of \ our \ pipe)$$

$$Q = \frac{4 \ A \ X}{\sqrt{y}}$$

$$Q = \frac{4 \times .22 \times 2}{\sqrt{3}}$$

$$Q = 1.02 \ cfs$$

$$1.02 \ cfs \times 60 \ sec./min. \times 1440 \ min./day \times 7.48 \ gal./cu.ft. = 660,000 \ gpd$$

20. **Answer**   4.3 inches

Assume hose is flowing full; change inches to feet; solve for diameter; change to inches.

$$18 \ inches = 1.5 \ ft.$$

$$Q = \frac{4 \ A \ X}{\sqrt{y}}$$

$$1 = \frac{4 \times .785 \ d^2 \times 3}{\sqrt{1.5}}$$

$$.36 \ ft. = d$$

$$4.3 \ inches = d$$

21. **Answer**   5.1 minutes

Draw diagram.   Calculate volume of water to be removed.   Solve for flow through the orifice; divide volume by flow to obtain time.

$$Vol. = 1 \times w \times h$$

$$Vol. = 12 \times 4 \times 3$$

Vol. = 144 cu.ft. This much water will be lost through the orifice.

$$Q = C_d \quad A \quad \sqrt{(Ph_1 - Ph_2) \times 2g}$$

$$Q = .6 \times .785 \times .25^2 \quad \sqrt{(4 - 0) \times 64.4}$$

$$Q = .47 \text{ cfs}$$

$$\frac{144 \text{ cu.ft.}}{.47 \text{ cu.ft./sec.}} = 306 \text{ sec.} = 5.1 \text{ minutes}$$

---

22. **Answer**   95 ft.

Draw diagram. Solve for head on weir; add to weir height to notch for depth of water behind weir; subtract the depth of flow upstream of backup; use slope to calculate length of backup.

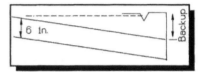

$$Q = 2.5 \quad h^{2.5}$$

$$2 = 2.5 \quad h^{2.5}$$

$$.9 \text{ ft.} = h$$

1.5 ft. weir ht. to notch + .9 ft. head = 2.4 ft. depth behind weir.

2.4 ft. depth behind weir - .5 ft. flow depth = 1.9 ft. backup depth

$$\frac{1.9 \text{ ft. backup depth}}{.02 \text{ ft./ft. slope}} = 95 \text{ ft. backup}$$

---

23. **Answer**   1191.6 ft.

Draw diagram. Solve for pressure on gates; solve for flow through gates; total flow for 6 gates; calculate amount going over the top; solve for depth (h on weir); add to original surface elevation; make adjustment.

1184 ft. (elev. of surface)
1126 ft. (elev. of gate center)
  58 ft. depth of water - pressure on gate

$$Q = C_d \quad A \quad \sqrt{(Ph_1 - Ph_2) \times 2g}$$

$$Q = .85 \times (6 \times 8) \quad \sqrt{58 \times 64.4}$$

$$Q = 2494 \text{ cfs . Flow through one gate}$$

2494 cfs x 6 gates = 14964 cfs through all gates

30,000 cfs (total flow) - 14964 cfs (flow through gates) =
15036 cfs over top

$$Q = 3.33 \quad L \quad h^{1.5}$$

$$15036 = 3.33 \times 200 \times h^{1.5}$$

$$8 \text{ ft.} = h$$

1184 ft. (elev. of dam top) + 8 ft. head = 1192 ft. (elev. of water going over dam)

Stop! Once dam is in place and there is a head over it, the pressure on the gates is greater - and the flow through the gates is greater. Previous calculation is not accurate. Many adjustments would be needed for precision, but let's do it once.

```
  1192 ft. (new top surface elevation)
- 1126 ft. (center of gate)
    66 ft. depth over gate
```

$$Q = C_d \quad A \quad \sqrt{(Ph_1 - Ph_2) \times 2g}$$

$$Q = .85 \times 48 \quad \sqrt{66 \times 64.4}$$

$$Q = 2660 \text{ cfs}$$

2660 ft. x 6 gates = 15960 cfs through gates

30,000 - 15960 cfs = 14040 cfs over top

$$Q = 3.33 \quad L \quad h^{1.5}$$

$$14040 = 3.33 \times 200 \times h^{1.5}$$

$$7.6 \text{ ft.} = h$$

1184 ft. (dam top) + 7.6 ft. head = 1191.6 ft. water surface over dam

24. **Answer**   4.89 ft./sec.

Solve for velocity; adjust for head loss.

Differential pressure = 3.5 inches = .3 ft. of water

Pitot tube reads Pressure Head at one gage and Pressure Head plus Velocity Head at the other. Differential pressure reading on manometer is equivalent to Velocity Head.

$$Vh = \frac{V^2}{2g}$$

$$.3 = \frac{V^2}{64.4}$$

$$4.4 \text{ ft./sec.} = V$$

Coefficient of discharge is a decimal number designating the % of loss of energy because of friction upon entering the unit. In the absence of a formula for pitot tubes, we can still use it by adjusting the velocity.

$$\frac{4.4 \text{ ft./sec.}}{.9} = 4.89 \text{ ft./sec.}$$

---

25. **Answer**   1.7 ft.

Draw diagram.   Calculate movement of mercury; subtract from top.

$$\frac{2 \text{ ft. of water pressure}}{13.6 \text{ (SG mercury)}} = .147 \text{ ft. (mercury moves this much up on right side)}$$

Before water was poured in, mercury rested at the zero point on both sides. On left side, water fills the tube from the top down to .147 ft. below the zero mark (because it moved the mercury out by this much). If the water extends to .147 ft. below the zero, and the mercury rose to .147 ft. above the zero, then distance to the top of the tube on the right is:

$$2 - .147 - .147 = 1.7 \text{ ft. to top of tube}$$

---

26. **Answer**   2.28 MGD

Method #1

$$Q_f = Q_t \left(\frac{P_s - P_f}{P_s - P_t}\right)^{.54}$$

$$Q_f = 1000 \left(\frac{60 - 20}{60 - 43}\right)^{.54}$$

$$Q_f = 1587 \text{ gpm}$$

$$Q_f = 2.28 \text{ MGD}$$

Method #2

Head loss under fire conditions is 60 psi - 20 psi = 40 psi = 92.4 ft.
Slope for this head loss over 300 ft. is .308 ft./ft.

Head loss for test is 60 psi - 43 psi = 17 psi = 39.3 ft.
Slope for this head loss over 300 ft. of test space is .1309 ft./ft.

Test flow is 1000 gpm = 2.232 cfs

$$\frac{Q_f}{s_f^{.54}} = \frac{Q_t}{s_t^{.54}}$$

$$\frac{Q_f}{.308^{.54}} = \frac{2.232}{.1309^{.54}}$$

$$Q_f = 3.5 \text{ cfs}$$

$$Q_f = 2.28 \text{ MGD}$$

chapter 11

1. **Answer** 130.2 gpd/capita

$$\frac{5,207,600 \text{ gal.}}{40,000 \text{ people}} = 130.2 \text{ gpd/capita}$$

2. **Answer** $16,060

$$\frac{200000 \text{ gpd}}{7.48 \text{ gal./cu.ft.}} = 26738 \text{ cu.ft./day used}$$

8% lost revenue; 92% billed

26738 cu.ft./day x .92 = 24599 cu.ft./day billed.

Calculate actual revenue.
$$\frac{24599 \text{ cu.ft./day x \$20.00}}{1000} = \$490/\text{day x 365 days/yr.} = \$178,850/\text{yr.}$$
actualrevenue

Calculate correct revenue.
$$\frac{26738 \text{ cu.ft./day x \$20.00}}{1000} = \$534/\text{day x 365 days/yr.} = \$194,910/\text{yr.}$$
correct revenue

$ 194910 yr.
<u>- 178850</u> yr.
$16060 yr. lost annually

3. **Answer** $346.52

Solve for gal./day used by let run; change to cu.ft.; solve for cost.

2 gal./min. x 1440 min./day x 30 days/mo. divided by 7.48 gal./cu.ft. = 11551 cu.ft./mo. water used

11551 cu.ft./mo. x $.03/cu.ft. = $346.52 deducted from bill.

4.   **Answer**   6.8%

Wastewater treatment plant flow should be the same as the drawdown of the water tower. The difference between them is lost water.

Water Tower Drawdown:

Vol. = .785 d^2 x h

Vol. = .785 x 50^2 x 5

Vol. = 9812.5 cu.ft.

Vol. = 73398 gal./3 hrs.

Wastewater treatment plant:

380 gal./min. x 60 min./hr. x 3 hrs. = 68400 gal./3 hours

```
  73398 gal. water made
- 68400 gal. water used
   4998 gal. lost water
```

$$\frac{4998 \text{ gal.}}{73398 \text{ gal.}} = 6.8\% \text{ lost water}$$

---

5.   **Answer**   24,615

2,000,000 gal./day x .2 = 400,000 gal./day lost

```
  2,000,000 gal./day made
-   400,000 gal./day lost
  1,600,000 gal./day available for the public
```

$$\frac{1,600,000 \text{ gal./day}}{65 \text{ gal./day/person}} = 24,615 \text{ people}$$

---

6.   **Answer**   625 miles per $1 water

Calculate gal. water needed to make 1 gal. gas. Driver gets 25 miles with this gas. Calculate gal. water purchased for one dollar. Convert to miles per dollar of water.

$$\frac{1000 \text{ gal. gas}}{20,000 \text{ gal. water}} = \frac{1 \text{ gal. gas}}{14 \text{ gal. water}}$$

20 = gal. water/1 gal. gas

Driver gets 25 miles with this 1 gal. gas, made with 20 gal. water.

Water costs $2.00/1000 gal. or $1/500 gal.

$$\frac{20 \text{ gal. water}}{25 \text{ miles}} = \frac{500 \text{ gal. water}}{15 \text{ miles}}$$

x = 625 miles/500 gal. water ($1)

7. **Answer**   43% of revenue

Let x = water use per customer service connection

Regular Customers		New Customers	Total Water Use
800 x	+	200 x =	400,000

x = 400 gpd/customer service connection

Regular Customers
400 gpd/customer x 800 customers = 3200,000 gpd used

$\frac{3200 \text{ gpd}}{1000}$ x $1.50 = $480/day x 365 days/yr. = $175,200/yr.

New Customers
400 gpd/customer x 200 customers = 80,000 gpd used

$\frac{80,000 \text{ gpd}}{1000}$ x $4.50 = $360/day x 365 days/yr. = $131,400/yr.

$175200/yr. revenue (regular customers)
+ $131400/yr. revenue (new customers)
$306600/yr. total revenue

$\frac{\$131,400 \text{ yr. new}}{\$306,600 \text{ yr. total}}$ = .429 = 43%

8. **Answer**   5.7% on the small side.

858200 gal. (turbine)
+   52000 gal. (pos. displ.)
910200 gal. total

$\frac{5200 \text{ gal. (pos. displ.}}{910200 \text{ gal. (total)}}$ = .057 = 5.7%

Less than 6% of flow on the small side. Many utilities would not consider this worth the capital cost and maintenance of a compound meter. The newer turbo meters are good at picking up small flows, and some of this may not be lost. However, it is a matter of choice.

9. **Answer** $2120.00

Change meter reading to gallons; total up cost of gallons used at each rate; add.

500,000 cu.ft. x 7.48 gal./cu.ft. = 3,740,000 gallons used

First 200,000 gallons:
@ $1.00/1000 gal. = $1.00 x 200 = $200.00

Next 500,000 gallons:
@ $.80/1000 gal. = .80 x 500 = $400.00

3,740,000 tot. gallons used
-   700,000 gallons already totalled
3,040,000 gallons left to be billed at $.50/1000

3,040,000 gal. (@$.50/1000 gal.) = $.50 x 3040 = $1520.00

Adding: $1,520.00 + $400.00 + $200.00 = $2,120.00 paid for water this
month.

---

10. **Answer**   A. 26.2%
               B. 13%

A. Add up accounted-for water; subtract from water made; obtain percentage.

fire water:
2000 gpm x 60 min.hr. x 141 hr./yr. = 16,920,000 gal./yr.

16,920,000 gal./yr. div. by 365 days/yr. = 46,356 gal./day = .05 MGD

underregistering meters:
8.71 MGD (customer meters) + .37 MGD (public meters) = 9.08 MGD

9.08 x .98 = 8.9 MGD actual metered water

8.90 MGD metered water
+ .05 MGD fire water
8.95 MGD accounted for

$\dfrac{\text{8.95 MGD accounted for}}{\text{12.12 MGD total made}}$ = 73.8% accounted for   26.2% Lost Water

B.   8.90 MGD metered water
          1.60 MGD unmetered public
     + .05 MGD fire
         10.55 MGD accounted for

$\dfrac{\text{10.55 MGD accounted for}}{\text{12.12 MGD total made}}$ = 87% accounted for   13% Lost Water

Lost Water taken as water lost through leaks, not water lost for revenue.

---

11. **Answer**   286.8 gpm

Consider only nighttime use.

Water storage tank fills at rate of 2 ft./hr. Find gpm filling rate.

$$Vol. = .785 \times d^2 \times h$$

$$Vol. = .785 \times 40^2 \times 2$$

$$Vol. = 2512 \text{ cu. ft.}$$

$$Vol. = 18789.8 \text{ gal.}$$

Tank fills at a rate of 18789.8 gal./hr.

This is 313.2 gpm.

Water tank:
Rate pumped in - industry use - domestic use = Rate of fill

1200 gpm - 600 gpm - x = 313.2 gpm

x = 286.8 gpm Domestic Use

---

12. **Answer**   10 inch diameter

Pressure increases from main to industry because of pipe incline; therefore the new service line can have 30 ft. HL - no more. Use this as allowable HL, solve for slope, then diameter.

$$3.8 \text{ MGD} \times 1.55 \text{ cfs/MGD} = 5.9 \text{ cfs}$$

$$Slope = \frac{head\ loss}{length} = \frac{30\ ft.}{1000\ ft.} = .03\ ft./ft.$$

$$Q = .435\ C\ d^{2.63}\ s^{.54}$$

$$5.9 = .435\ 150\ d^{2.63}\ .03^{.5}$$

$$.82 \text{ ft.} = d$$

$$10 \text{ in.} = d$$

---

13. **Answer**   2 in. pos. displ. plus 6 in. turbine

Trailer Court

Limited use first few years plus senior vacancy in winter will mean low flows must be provided for. This is continual use, and revenue.

Fire flow needed at all times = 1000 gpm; only occasional use at the flow.

Domestic flow when full:
From Fixture Value Chart:
  Bathtub . . . .   8 gpm
  Shower . . . . 4 gpm
Water Closet . .   3 gpm
Dishwasher . . .   4 gpm
Washing Machine   5 gpm
Kitchen Sink . . .   <u>3 gpm</u>
              27 gpm x 130 trailers = 3510 gpm total

Adjusted value from Water-Flow demand chart = 3510 gpm, high range = 80 gpm

Full park domestic flow: 80 gpm

Present flow = 27 gpm -- adjusted 18 gpm daily use (less at night)

Fire flow = 1000 gpm

No one meter will service such high and low flows. A large meter would pass too much of the low flows. Install two meters - a 6 inch turbine on the 6 inch line, with a spring loaded check valve downstream which will not open until fire flows are encountered. On a bypass install a 2 inch positive displacement meter (records flows from 1.25 to 160 gpm) which will be in use most of the time.

Note was made of winter vacancies. (cold weather). Install both meters in a pit to prevent freezing.

---

14. **Answer**   2 in. pos. displ. or 2 inch compound

Ice Cube Factory

Flow required:
Ice Cube Machines (2) . . . . . .   30 gpm
Compressors . . . .   25 gpm

Domestic:
  Sink . . . .   4 gpm
  Water Closet . . . .   3 gpm
  Hose Connections (4) . . . .   <u>+ 24 gpm</u>
                             31 gpm

Domestic adjusted value = 18 gpm

Total flow required with expansion:
30 gpm + 25 gpm + 18 gpm = 73 gpm

Ice cube machines will be on during the day only, as will domestic use, but compressors will be 24 hr./day use - lower flow at night.

Could use 2 inch positive displacement (records 8 to 160 gpm) or 2 inch compound meter (records 1 to 160 gpm).

Compound is more expensive, but catches all the low flows.

---

15. **Answer**   ?

Golf Club Irrigation System

RPPBP was needed because he already has booster pumping installed and intends to use when pressure is too low to get adequate water to far ends of course.

RP will drop pressure by 15-20 psi.

He wants a meter with low head loss. This knocks out the use of pos. displ., which would only be good for the low flows. Same with compound.-high head loss.

Turbine meter needs length of straight pipe upstream; this is not possible in this meter house.

This is an unusual situation. A right angle 6 inch propeller meter was installed (head loss = 1.5 ft.). Flow range 90-1200 gpm. This will not pick up the low flows; it will not register at all below 30 gpm.

The utility loses revenue.

What would you have installed?

---

16. **Answer**   2 inch diam. DIP service line - 2 inch pos. disp. meter

Apartment building - 20 units - obtain flow

From Fixture Value Chart:
Bathtub (20) @ 8 gpm . . . . .   160 gpm
Water Closet (40) @ 3 gpm . .   120 gpm
Shower (40) @ 4 gpm . . . . .   160 gpm
Sink (40) @ 4 gpm . . . .   160 gpm
Dishwasher (20) @ 4 gpm . . .   80 gpm
Kitchen Sink (20) @ 3 gpm . .   60 gpm
Hose Connections (4) @ 10 gpm   40 gpm
Washing Machines (4) @ 12 gpm 48 gpm
Service Sinks (2) @ 7 gpm   + 14 gpm
                                842 total gpm

Adjust with Water-Flow Demand Chart (low range) -- 60 gpm practical flow needed

Draw diagram of system:

<u>Will need 60 gpm (.133 cfs) at 35 psi.</u>

Elevation difference: Assume main 6 ft. below ground; allow 10 ft. different for elevation (4.3 psi).

No RP device needed.

<u>Choose 2 inch diameter DIP (.167 ft. diam.) C = 140</u>

Fittings:
3 open gate valves . .   4.5 ft. equiv. pipe
2 std. elbows . . . . .   10 ft. equiv. pipe
straight pipe . . . .   <u>+ 300 ft.</u>
                    315 ft. total pipe length with fittings

Solve for HL$_f$:
$Q = .435 \; C \; d^{2/63} \; s^{.54}$

$.133 = .435 \quad 140 \quad .167^{2.63} \; s^{.54}$

$.0722 = s$

.0722 ft./ft. x 315 ft. pipe $= 22.7$ ft. HL$_f$ $= 9.9$ psi

80 psi at main - 4.3 psi (elev.) - 9.9 psi (HL$_f$) $= 65.8$ psi left at meter.

Check velocity at this flow:
$Q = A \; V$

$.133 = .785 \; x \; .167^2 \; V$

6 ft./sec. $= V$

Velocity OK.

Continue. Choose a meter for the installation.

From Meters By Flow chart:
<u>2 inch positive displacement</u> will service up to 80 gpm max continuous demand.

This meter loses about 11 psi through the meter at 60 gpm.

Pressure 65.8 psi - 11 psi $= 54.8$ psi still left. Plenty of pressure, allows for some drop inside building.

17. **Answer**   6 inch service - 4 inch class II turbine

Required 200 gpm

Expansion 200 gpm - in Florida community it may easily double production.

> Bathroom: (from fixture value table)
> Sink . . . . 4 gpm
> Water Closet . . . . 3 gpm
>
> Also needed:
> Safety Shower . . . . 4 gpm
> Eyewash . . . . 2 gpm
> Service Sink . . . . $\underline{+\ 7\ gpm}$
> 20 total gpm

Adjust from Water-Flow Demand per Fixture Value Chart: Yields 15 gpm for this domestic use.

400 gpm industrial use + 15 gpm domestic use = <u>415 gpm Total flow needed</u> (.927 cfs)

Industry intends to install 1½ inch interior piping (.125 ft.)

Try this size for service piping. Check velocity.

$Q = A \quad V$

$.927 = .785 \times .125^2 \quad V$

76 ft./sec. = V

No good - way too fast.

NOTE: Interior piping - if flow is not split up immediately upon entrance, 1½ inch piping is too small inhouse. Lots of surge control and antisplash devices needed. Head loss through 50-80 ft. of this would be over 600 ft.

> **Service Pipe:**
> Try 3½ inch pipe (.292 ft.) Check velocity.
>
> $Q = A \quad V$
>
> $.927 = .785 \times .292^2 \quad V$
>
> 14 ft./sec. = V
>
> This will handle expansion flow and still have decent velocity. OK!
>
> Draw diagram of system:

Fittings:

5 open gate valves . . .	10 ft. equiv. length
2 std. elbow	10 ft. equiv. length
Pipe length	+ 200 ft.
	220 ft. total length

65 psi at main

Elevation difference = 4 ft. (Fla. piping 1-2 ft. below ground) = 1.7 psi

RP device needed (electroplating) = 25 psi drop across it

Head Loss in service pipe: Choose new PVC: C=150
$$Q = .435 \quad C \quad d^{2.63} \quad s^{.54}$$

$$.927 = .435 \quad 150 \quad .292^{2.63} \quad s^{.54}$$

$$.153 = s$$

.153 ft./ft. x 220 ft. = 33.7 ft. HL_f = 14.6 psi

Subtract pressure losses:
65 psi - 1.7 psi (elev.) - 25 psi (RP device) - 14.6 psi (HL_f) = 23.7 psi at meter. Not enough pressure!

- This does not include any inhouse piping losses.
- It does not include the meter.
- Even at present flow of 215 gpm, the HL_f would be 10 ft., and pressure at the meter would drop to 34 psi - acceptable, but allows for no expansion.

A 6 inch diameter service pipe could be installed. HL_f would be:
$$Q = .435 \quad C \quad d^{2.63} \quad s^{.54}$$

$$.927 = .435 \quad 150 \quad .5^{2.63} \quad s^{.54}$$

$$.011 = s$$

.011 ft./ft. x 220 ft. = 2.4 ft. HL_f = 1 psi

Again: 65 psi - 1.7 psi - 25 psi - 1 psi = 37 psi at meter.

Choose a meter:
From Meters By Flow chart, a 4 inch turbo meter would do. Max continuous use is 1000 gpm. Head loss is very low in turbo meters. Use would be during day, and at max. capacity all the time (no low flows).

This industry will probably still need booster pumping inhouse.

An alternate would be for them to have their own elevated water storage, a secondary supply, separated from community supply by air gap.

chapter 12

1. **Answer**   A. 9 hp
            B. 100 ft.
            C. 120 ft.
            D. 118 ft. and 46 ft.

To avoid mistakes, work from the flow.

A. On left, locate 90 ft. head; follow across to H-Q curve, then down to flow (276 gpm); now go back up to Hp curve, and across to the right to read hp = 9 hp.

B. Locate top of E-Q curve; slide down to the flow (240 gpm), then back up to the H-Q curve, then left to head = 100 ft.

C. Top of H-Q curve is shutoff head. Pump can lift the water only this high. Shutoff head = 100 ft.

D. This pump will be 30% efficient at two different heads.
   Locate 30% efficiency on left side of E-Q curve; slide down to flow (62 gpm); move straight up to H-Q curve, and read head on left = 118 ft. Now go to right side of E-Q curve (will have to extend the curves down a little to meet this point); slide down to flow (395 gpm), then back up to H-Q curve, and left to read head = 46 ft.

Note that if head is very high, it pumps little water and is very inefficient. Larger pump should have been purchased. If head is very low, a great deal of water is pumped, but also at low efficiency, for horsepower requirements are high.

2. **Answer**   A. 90 ft. & 155 ft.
            B. 1500 gpm
            C. 45 ft.

A. Locate E-Q curve; find point of 70% efficiency; slide down to flow, then up to H-Q curve; look left to TDH = 90 ft. Back to E-Q curve; find 70% efficiency on right side, slide down to flow, up to H-Q curve, over to head = 155 ft.

B. Locate head, slide down to flow = 1500 gpm

C. Locate new flow, up to H-Q curve, over to head = 45 ft. or less. Sudden great increase in flow and drop in discharge pressure indicates break in discharge line.

3. **Answer**   A. 240 gpm
                  B. 0
                  C. 0

For each, draw diagram, solve for head; locate flow on curve.

A.   TDH = Lift + Losses

    TDH = (100 - 25) + (15 + 10)

    TDH = 100 ft.

    At this head the pump is operating at 240 gpm, the design point.

B.   TDH = Lift + Losses

    TDH = (100 + 15) + 35 + 5)

    TDH = 155 ft.

    Head required is way over shutoff head. No flow delivered.

C.   TDH = Lift + Losses

    TDH = (50 + 20) + (12 + 13)

    TDH = 95 ft.

    At 95 ft. of head pump delivers 260 gpm.

    Theoretically flow is 260 gpm. However, unless this is a specialized well pump, suction lift and losses are too great; water would vaporize in the line before it got up to the pump. There would be no flow.

4. **Answer**   103 ft.

Draw Diagram. Referring to curve, TDH at 200 gpm = 108 ft.; losses are 20 ft.

          TDH = Lift + Losses

          108 = x + 20

          88 ft. = x

          88 ft. lift + 15 ft. static suction head = 103 ft. up to end of discharge.

5. **Answer**   A. 80%
                 B. 7.5 mhp
                 C. 95 ft.

A. Refer to top of E-Q curve; look right for efficiency = 80%.

B. To find bhp, locate flow (140 gpm) on curve; slide up to P-Q curve; read bhp on right = 6 hp; divide by motor efficiency.

$$mhp = \frac{bhp}{motor\ eff.} = \frac{6}{.8} = 7.5\ mhp$$

C. Find 8 hp on right; slide left to P-Q curve, then down to flow (260 gpm); At 260 gpm this pump uses 8 hp. Slide up from flow to H-Q curve; then left to head = 95 ft.

---

6. **Answer** $7.85

Draw Diagram. Solve for total head; then whp, then mhp, then cost.

TDH = Lift + Losses

TDH = (50 + 15) + 12

TDH = 77 ft.

$$\frac{gpm\ x\ hd}{3960} = whp$$

$$\frac{100\ x\ 77}{3960} = whp$$

1.9 hp = whp

wire to water eff. = .76 x .85 = .65

$$\frac{1.9\ whp}{.65} = 2.9\ mhp$$

2.9 mhp x .746 Kw/hp x 24 hr./day x $.15 Kw/hr. = $7.85/day

---

7. **Answer**   1719 ft.

Solve for whp; solve for flow; solve for head; solve for lift.

$$\frac{60\ Kw\ at\ motor}{.746\ Kw/Hp} = 80\ mhp$$

motor hp x wire to water eff. = whp

80 x .6 = 48 hp

$$Q = A \quad V$$

$$Q = .785 \text{ x } .25^2 \text{ x } 5$$

$$Q = .245 \text{ cfs}$$

$$Q = 110 \text{ gpm}$$

$$\frac{\text{gpm x hd}}{3960} = \text{whp}$$

$$\frac{110 \text{ x hd}}{3960} = 48$$

1728 ft. = hd. This is TDH (lift plus losses).

To obtain lift, subtract losses: 1728 ft. - 9 ft. = 1719 ft. lift.

---

8. **Answer**   98 ft. and 1200 gpm (84% efficiency)

Graph paper - along bottom set up range of flow (200-1600 gpm)
Up the side set up heads - up to 140 ft. at top, in 20 ft. increments.
For each flow stated, place dot at the corresponding head; draw H-Q curve.

Up right side set up efficiencies - up to 100%.
Plot each flow against efficiency the same way; draw E-Q curve.

On either side separately, set up horsepowers; calculate brake hp for each head,
flow and efficiency. Plot each flow against its bhp on graph.

Choose the greatest efficiency this pump achieves; locate head and flow; this is the
design point.

---

9. For each flow, head and efficiency, calculate bhp; plot this against flow on the
graph; set up hp range on either side.

---

10. **Answer**   86%

Calculate mhp; refer to curve for head and calculate bhp; calculate motor eff.

$$\text{mph x } .746 \text{ Kw/hp x hr./day x days/mo. x rate} = \text{cost}$$

$$\text{mph x } .746 \text{ x } 24 \text{ x } 30 \text{ x } .10 = 5000$$

$$\text{mhp} = 93 \text{ hp}$$

$$\frac{\text{gpm x hd}}{3960 \text{ x pump eff.}} = \text{bhp}$$

$$\frac{800 \text{ x } 276}{3960 \text{ x } .7} = \text{bhp}$$

$$80 \text{ hp} = \text{bhp}$$

$$\frac{bhp}{mhp} = \text{motor eff.}$$

$$\frac{80}{93} = 86\% \text{ efficient}$$

---

11. **Answer** $5,890.00

Add lift to Losses for TDH; refer to curve to calculate mhp; calculate mhp; calculate cost.

240 ft. lift + 27 ft. HL = 267 ft. TDH

$$\frac{\text{gpm x head}}{3960 \text{ x pump eff.}} = bhp \qquad\qquad mhp = \frac{bhp}{\text{motor eff.}}$$

$$\frac{860 \times 267}{3960 \times .47} = bhp \qquad\qquad mhp = \frac{123.4}{.9}$$

123.4 hp = bhp   mhp = 137 hp

137 mhp x .746 Kw/hp x 24 hr./day x 30 days/mo. x $.08 = $5,890.00

---

12. **Answer**   A. 120 ft.
                           B. 325 gpm

A. At shutoff head it draws only 6 hp. It is also not pumping any water.

B. Solve for head; refer to curve for flow.

TDH = Lift + Losses

TDH = 12 + (.015 x 4000)

TDH = 72 ft.

At 72 ft. of head, the pump will deliver 325 gpm.

---

13. **Answer**   A. 41 psi
                           B. 9.1 hp

Draw diagram.

A. In order to solve for tot. hd.; solve for flow; solve for suction and discharge head losses; add lift + losses for TDH.

Q = A V

Q = .785 x .33² x 6

Q = .513 cfs

Suction:

$Q = .435 \quad C \quad d^{2.63} \quad s^{.54}$

$.513 = .435 \quad 100 \quad .33^{2.63} \quad s^{.54}$

$.059 = s$

.059 ft./ft. x 20 ft. = 1.2 ft. head loss on suction side

Discharge:

$Q = .435 \quad C \quad d^{2.63} \quad s^{.54}$

$.513 = .435 \quad 100 \quad .25^{2.63} \quad s^{.54}$

$.229 = s$

.229 ft./ft. x 100 ft. = 23 ft. head loss on discharge side

TDH = Lift + Losses

TDH = 70 + (1.2 + 23)

TDH = 94.2 ft. head

$$\frac{94.2 \text{ ft.}}{2.31 \text{ ft./psi}} = 41 \text{ psi}$$

B. Convert flow to gpm; solve for hp.

.513 cfs x 60 x 7.48 = 230 gpm

$$\frac{\text{gpm x head}}{3960 \times \text{pump eff.}} = \text{bhp}$$

$$\frac{230 \times 94.2}{3960 \times .6} = 9.1 \text{ hp}$$

NOTE: Choice of C factor varies headloss.

---

14. **Answer**   A. 91 gpm
                  B. 91%

A. Obtain volume; solve for pumping rate.

Vol. = 1 x w x h x 7.48 gal./cu.ft.

Vol. = 150 x 10 x 35 x 7.48

Vol. = 392700 gallons

1440 min./day x 3 days = 4320 min. pumping time

$$\frac{392700 \text{ gallons}}{4320 \text{ min.}} = 91 \text{ gpm}$$

B. Divide by rated pumping rate.

$$\text{Eff} = \frac{91 \text{ gpm actual rate}}{100 \text{ gpm rated rate}} = .91 = 91\% \text{ efficient}$$

---

15. **Answer**   115 ft.

Solve for friction loss in force main; add to minor loss and static discharge head for TDH.

600 gpm x 1440 min./day = 864000 gpd = .864 MGD

.864 MGD x 1.55 cfs/MGD - 1.34 cfs

$Q = .435 \quad C \quad d^{2.63} \quad s^{.54}$

$1.34 = .435 \quad 120 \quad .67^{2.63} \quad s^{.54}$

$.008 = s$

.008 ft./ft. x 4600 ft. = 37 ft. friction loss

TDH = Lift + Losses

TDH = 73 + (37 + 5)

TDH = 115 ft.

---

16. **Answer**   6.4 pipes

Choose length and C factor; obtain head loss through 12 inch pipe at this flow; now obtain flow for a 6 inch pipe with the same characteristics; solve for number of pipes needed.

One foot of pipe chosen; C = 100

1500 gpm = 3.34 cfs

$Q = .435 \quad C \quad d^{2.63} \quad s^{.54}$

$3.34 = .435 \quad 100 \quad 1^{2.63} \quad s^{.54}$

$.008 = s$

$Q = .435 \quad C \quad d^{2.63} \quad s^{.54}$

$Q = .435 \quad 100 \quad .5^{2.63} \quad .008^{.54}$

Q = .52 cfs will pass through a 6 inch pipe

$$\frac{3.34 \text{ cfs}}{.52 \text{ cfs/pipe}} = 6.4 \text{ pipes}$$

17. **Answer**   2.6 hp

This pump is only required to provide for head losses. There is no lift.
Convert fittings to length of straight pipe, add to pipe length; calculate slope, then
head loss; calculate whp.

> **Suction line:**
> | | |
> |---|---|
> | Length | 60 ft. |
> | 1 gate open | 2 ft. |
> | 4 90 degree ls els | 28 ft. |
> | d/D ½ | + 4 ft. |
> | | 94 ft. |

> **Discharge line:**
> | | |
> |---|---|
> | Length | 1800 ft. |
> | 1 gate open | 2 ft. |
> | 1 globe open | 100 ft. |
> | 3 90 degree ls els | 21 ft. |
> | d/D ½ | 6 ft. |
> | 4 tees | + 28 ft. |
> | | 1957 ft. |

Total pipe length: 94 ft. + 1957 ft. = 2051 ft.

200 gpm = .45 cfs; new DIP C = 140

$$Q = .435 \quad C \quad d^{2.63} \quad s^{.54}$$

$$.45 = .435 \quad 140 \quad .33^{2.63} \quad s^{.54}$$

$$.025 = s$$

.025 ft./ft. x 2051 ft. = 51.3 ft. head loss in piping

Pumping head = 51.3 ft.

$$\frac{gpm \times hd}{3960} = whp$$

$$\frac{200 \times 51.3}{3960} = 2.6 \ hp$$

---

18. **Answer**   750 gpm

Draw Diagram. Must create system curve; obtain head
at a range of flows; where the H-Q curve of #5
intersects, this is the operating point.

TDH = Lift + Losses

Lift = 15 ft. suction lift + 250 ft. discharge
lift = 265 ft. lift.

Losses = HL for pipe length + fittings

Convert fittings to pipe length:

Suction side:

Straight pipe	30 ft.
Borda entrance	30 ft.
Std. elbow	25 ft.
1 gate open	+ 6 ft.
	91 ft.

Discharge side:

Straight pipe	5000 ft.
1 check	60 ft.
1 gate open	6 ft.
2 std. els	50 ft.
1 ord. exit	+ 30 ft.
	5146 ft.

91 ft. + 5146 ft. = 5237 ft. total length of pipe

Try 600 gpm (1.34 cfs)   New steel   C = 140

$$Q = .435 \quad C \quad d^{2.63} \quad s^{.54}$$

$$1.34 = .435 \quad 140 \quad .83^{2.63} \quad s^{.54}$$

$$.002 = s$$

.002 ft./ft. x 5237 ft. pipe = 10.5 ft. HL

TDH = 265 ft. lift + 10.5 ft. loss = 275.5 ft. TDH

On curve #5, put an x at 600 gpm and 275.5 ft.

Try 800 gpm (1.78 cfs).

$$Q = .435 \quad C \quad d^{2.63} \quad s^{.54}$$

$$1.78 = .435 \quad 140 \quad .83^{2.63} \quad s^{.54}$$

$$.004 = s$$

.004 ft./ft. x 5231 ft. pipe = 20.9 ft. HL

TDH = 265 ft. lift + 20.9 ft. loss = 285.9 ft. TDH

Put an x at 800 gpm and 286 ft.

Try 1000 gpm (2.23 cfs).

$$Q = .435 \quad C \quad d^{2.63} \quad s^{.54}$$

$$2.23 = .435 \quad 140 \quad .83^{2.63} \quad s^{.54}$$

$$.005 = s$$

.005 ft./ft. x 5237 ft. pipe = 26.2 ft. HL

TDH = 265 ft. lift + 26.2 ft. loss = 291 ft. TDH

Put an x at 1000 gpm and 291 ft. head.

Draw system curve.

Pump curve crosses system curve at 750 gpm. This is the operating point of this pump.

---

19. **Answer**   350 gpm

Draw Diagram. Develop system curve for the combined pump system.

Flow will depend on lift and losses.

Flows are additive. TDH = Lift (10 ft.) + (loss in small line + loss in big pipe).

Obtain head for this pump at minimal flow; (get HL in small and large line; make sure to add flow together in large line). Try different larger flows till this system curve crosses the pump curve.

> **Try 50 gpm from pump through small line:**
> (.11 cfs) C = 120
>
> $Q = .435$ C $d^{2.63}$ $s^{.54}$
>
> $.11 = .435$ 120 $.33^{2.63}$ $s^{.54}$
>
> $.002 = s$
>
> .002 ft./ft. x 210 ft. pipe = .5 ft. HL in small line
>
> **Loss in large line:**
> Q = 200 gpm + 50 gpm = .56 cfs (C=100)
>
> $Q = .435$ C $d^{2.63}$ $s^{.54}$
>
> $.56 = .435$ 100 $.67^{2.63}$ $s^{.54}$
>
> $.002 = s$
>
> .002 ft./ft. x 3000 ft. = 6 ft. HL
>
> Add head losses: 6 ft. + .5 ft. = 6.5 ft. Tot. HL if pump yields 50 gpm.
>
> TDH = 10 ft. suction lift + 6.5 ft. HL = 16.5 ft. TDH
>
> On curve #1, put an x at 50 gpm and 16.5 ft. head.
>
> Try 200 gpm from pump through small line (.45 cfs).

$Q = .435 \quad C \quad d^{2.63} \quad s^{.54}$

$.45 = .435 \quad 120 \quad .33^{2.63} \quad s^{.54}$

$.033 = s$

.033 ft./ft. x 210 ft. pipe = 7 ft. HL in small line

Loss in large line: Q = 200 gpm + 200 gpm = 400 gpm (.89 cfs)

$Q = .435 \quad C \quad d^{2.63} \quad s^{.54}$

$.89 = .435 \quad 100 \quad .67^{2.63} \quad s^{.54}$

$.005 = s$

.005 ft./ft. x 3000 ft. = 15 ft. HL

Add head losses: 7 ft. + 15 ft. = 22 ft. Tot. HL if pump yields 200 gpm.

TDH = 10 ft. suction lift + 22 ft. HL = 32 ft. TDH

On the curve, put an x at 200 gpm and 32 ft. head.

Try 350 gpm from pump through small line (.78 cfs).

$Q = .435 \quad C \quad d^{2.63} \quad s^{.54}$

$.78 = .435 \quad 120 \quad .33^{2.63} \quad s^{.54}$

$.09 = s$

.09 ft. x 210 ft. = 19.3 ft. HL in small line

Loss in large line: Q = 350 gpm + 200 gpm = 550 gpm (1.23 cfs)

$Q = .435 \quad C \quad d^{2.63} \quad s^{.54}$

$1.23 = .435 \quad 100 \quad .67^{2.63} \quad s^{.54}$

$.0095 = s$

.0095 ft./ft. x 3000 ft. = 28.6 ft. HL

Add head losses: 19.3 ft. + 28.6 ft. = 48 ft. HL if pump yields 350 gpm.

TDH = 10 ft. suction lift = 48 ft. HL = 58 ft. TDH

On curve, put an x on 350 gpm and 58 ft. head.

System curve crosses pump curve here; no need to go any further. This is the operating point of the pump in this system.

NOTE: the lack of efficiency at this flow; this pump was a poor choice for this system.

20. **Answer**   8 inch diameter

Draw Diagram.   Refer to curve for Tot. Head; subtract for head loss; solve for diameter.

From curve TDH = 158 ft.

150 ft. = Lift

8 ft. = Loss

$\dfrac{8 \text{ ft. head loss}}{1000 \text{ ft. pipe}}$ = .008 ft./ft. slope

500 gpm = 1.1 cfs

DIP C = 130

$Q = .435 \quad C \quad d^{2.63} \quad s^{.54}$

$1.1 = .435 \quad 140 \quad d^{2.63} \quad .008^{.54}$

.585 ft. = d

7.1 in. = d

Choose 8 inch diameter pipe.

---

21. **Answer**   A. 1300 gpm
                  B. 2600 gpm

A. When pumps are installed in series, the head increases; flow stays the same. If TDH = 140 ft., Q = 1300 gpm.

B. Pumps installed in parallel draw from separate suction, and flows are addictive. Head is the TDH which corresponds to the flow - for each pump. At a head of 140 ft. the parallel system will produce a flow of 2600 gpm.

---

22. **Answer**   420 gpm

TDH = 1000 ft. to surface + 80 ft. above surface + Losses

Losses depend upon flow. Develop system curve.

Pump curve #3 represents one stage (1 impeller + bowl); this is a "pumps in series" problem (impellers in series). Heads are additive. Solve for head loss -at a range of flows. Where it crosses the pump curve, this is the operating point. Using three impellers will triple the head (at each flow, this pump will achieve three times the head of the same pump with one impeller). The three impellers for this system was necessary, for the lift alone requires it.

Solve for the head loss at each flow - it must be added to the lift to develop the system curve & TDH.

Try 300 gpm: (.67 cfs)   Assume C = 100.

$Q = .435$ C $d^{2.63}$ $s^{.54}$

$.67 = .435$ 100 $1^{2.63}$ $s^{.54}$

$.0004 = s$

.0004 ft./ft. x 1200 ft. pipe = .5 ft. HL

TDH = 1080 ft. + .5 ft. = 1080.5 ft. TDH

Try 500 gpm: (1.1 cfs)

$Q = .435$ C $d^{2.63}$ $s^{.54}$

$1.1 = .435$ 100 $1^{2.63}$ $s^{.54}$

$.0011 = s$

.0011 ft./ft. x 1200 ft. = 1.3 ft. HL

TDH = 1080 + 1.3 = 1081.3 ft. TDH

To work this out accurately, a pump characteristic curve for the 3 stage pump is constructed. Take each flow (100,200,300,400,500 gpm), triple the head each achieves, and place an x at that point. Connect them to make the curve. This is the curve for the 3 stage pump.

Enter system curve on the graph. At no flow (static lift only) the TDH is 1080. Add head losses for each increasing amount of flow, and enter TDH's for these flows.

Head losses in this system are negligible. The system curve is very flat, and almost the total head needed is because of lift. We can divide the lift by three, neglect the head loss, and just extend the lift out to where it crosses curve #3 (H-Q). To adapt for use of this curve (impeller), obtain TDH (Total lift + Losses, then divide by 3. This yields the TDH that each impeller is achieving.)

23. **Answer**   A. 185 ft.
                 B. 80 hp

Assume 1.5 MGD will be an adequate supply for the expanded population. Assume city terrain is level.

A. 80 psi x 2.31 ft./psi = 185 ft. (Height of water surface in tank when full).

B.   **Pump:**
                          1.5 MGD = 2.33 cfs = 1044 gpm

                          TDH = Lift + Losses

                          TDH = 185 + losses in main between treatment plant and tower.

                          Find losses:
                          2 miles = 10560 ft. of pipe

                          Try 10 inch diameter main, new DIP (C = 140).

                          $Q = .435 \quad C \quad d^{2.63} \quad s^{.54}$

                          $2.33 = .435 \quad 140 \quad .83^{2.63} \quad s^{.54}$

                          $.006 = s$

                          .006 ft./ft. x 10560 ft. = 63.4 ft. HL (looks acceptable)

                          TDH = 185 ft. lift + 63.4 ft. loss = 248 ft. TDH

                          Assume high lift pump will be new and 85% efficient.

                          $\dfrac{\text{gpm x head}}{3960 \text{ x pump eff.}} = Bhp$

                          $\dfrac{1044 \text{ x } 248}{3960 \text{ x } .85} = Bhp$

                          77 hp = Bhp

                          An 80 hp pump will be needed.

                          A smaller pump may be possible if a 12 inch diameter main is used. Cost
                          of pump would have to be weighed against cost of pipe.

---

24. **Answer**   A. 7 hp
                    B. #4

Draw diagram.
A. Choose 6 inch diameter line; new steel pipe C =
    140.   Calculate equiv. pipe length for fittings;
    calculate slope, then HL; add to lift.

| | Suction Side: |
| | Suction lift = 8 ft. |

Pipe length	40 ft.
1 gate valve	3 ft.
1 foot valve	40 ft.
1 std. elbow	17 ft.
d/D ½	6 ft.
Borda entrance	+ 18 ft.
	124 ft. Total pipe length Suction

Discharge Side:
Discharge lift = 33 ft.

Pipe length	910 ft.
d/D ½	10 ft.
2 std. elbows	34 ft.
1 check valve	30 ft.
1 gate valve	3 ft.
2 tees	+ 20 ft.
	1007 ft. Total pipe length Discharge

124 ft. (suction length) + 1007 (discharge length) = 1131 ft. total length

Flow = 400 gpm = .89 cfs

$Q = .435 \quad C \quad d^{2.63} \quad s^{.54}$

$.89 = .435 \quad 140 \quad .5^{2.63} \quad s^{.54}$

$.012 = s$

.012 ft./ft. x 1131 ft. = 13.6 ft. HL

TDH = Lift + Losses

TDH = 41 + 13.6

TDH = 55 ft.

Calculate horsepower. Assume operation close to design point - new pump 85% efficient.

$$\frac{\text{gpm x head}}{3960 \text{ x pump eff.}} = Bhp$$

$$\frac{400 \text{ x } 55}{3960 \text{ x } .85} = Bhp$$

6.5 = Bhp

Purchase a 7 hp pump whose design point is at this head and flow.

B. Inspect available curves.

#1 - would not provide enough flow at this head.

#2 - would have to have discharge valve closed almost to shutoff head in order to provide correct flow. Pump is too big. Power requirements too high.

#3 - This is a deep well pump and doesn't operate in this range at all.

#4 - This one has possibilities. At 53 ft. head it will deliver 400 gpm. However, efficiency is only about 60% at this point.

#5 - No good. Like #2, would have to be operating almost at shutoff head to provide this low a flow.

Will choose pump #4. Will have to keep head losses down in order to provide sufficient flow; 60% efficiency is poor, however. Would not purchase this pump as first choice.

25. **Answer**   Need two 13 hp pumps which pump 1 MGD each
              and two 30 hp pumps which pump 2 MGD each

Draw Diagram.

2 MGD = 1389 gpm (3.1 cfs)

4 MGD = 2778 gpm (for future) (6.3 cfs)

Must be able to handle maximum flow with largest pump out of service.

Will need 2 small pumps, 1 MGD each - for low flow.

Will need 2 large pumps 2 MGD each - for high flow.

Bubbler for level control; signal relay for any 1, 2, or 3 pumps on.

TDH = lift (60 ft.) + losses (in 700 ft. of pipe)

Calculate head loss in 700 ft. line under worst condition (4 MGD flow - future)

Choose 12 inch diameter pipe; C = 130.

With 4 MGD flow:
$Q = .435 \ C \ d^{2.63} \ s^{.54}$

$6.2 = .435 \quad 130 \quad 1^{2.63} \ s^{.54}$

$.0166 = s$

$.0166$ ft./ft. x 700 ft. = 12 ft. HL in line at 4 MGD flow

With 2 MGD flow:
$Q = .435 \ C \ d^{2.63} \ s^{.54}$

$3.1 = .435 \quad 130 \quad 1^{2.63} \ s^{.54}$

$.005 = s$

$.005$ ft./ft. x 700 ft. = 3.2 ft. HL in line at 2 MGD flow.

Calculate pump size: Assume new pump 85% efficient.

Small Pump:
(1 MGD); 694 gpm. TDH = 60 ft. lift + max 3 ft. loss = 63 ft.

$$\frac{\text{gpm x head}}{3960 \text{ x } .85} = \text{Bhp}$$

$$\frac{694 \text{ x } 63}{3960 \text{ x } .85} = 12.9 \text{ hp}$$

Large pump: (both together 4 MGD) 2778 gpm. TDH = 60 ft. lift + 12 ft. HL = 72 ft. Separately 2 MGD; 1389 gpm.

$$\frac{\text{gpm x head}}{3960 \text{ x } .85} = \text{Bhp}$$

$$\frac{1389 \text{ x } 72}{3960 \text{ x } .85} = 30 \text{ hp}$$

chapter 13

1. **Answer**   A. 107 ft./sec.
                B. 178 ft.

A. Change diameter to ft.; find distance traveled for each revolution of impeller; solve for distance traveled/sec.

14 inches = 1.17 ft.

C = 3.14 x diam.

C = 3.14 x 1.17

C = 3.67 ft. (distance a point at impeller tip travels per revolution.)

Velocity = rev./min. x ft./rev.

V = 1750 x 3.67

V = 6423 ft./min.

V = 107 ft./sec.

B. All the velocity head is changed to pressure head upon exiting pump.

$$Vh = \frac{V^2}{64.4}$$

$$Vh = \frac{107^2}{64.4} = 178 \text{ ft. of pressure created by pump}$$

2. **Answer**   A. 214 gpm
             B. 67.3 ft.

   A. Capacity:

$$\frac{\text{New RPM}}{\text{Old RPM}} \times Q = \text{New Q}$$

$$\frac{1250}{1750} \times 300 = \text{New Q}$$

$$214 \text{ gpm} = \text{New Q}$$

   B. Head: (57 psi = 132 ft.)

$$\left(\frac{\text{New RPM}}{\text{Old RPM}}\right)^2 \times \text{head} = \text{New Head}$$

$$\left(\frac{1250}{1750}\right)^2 \times 132 = \text{New Head}$$

$$67.3 \text{ ft.} = \text{New Head}$$

---

3. **Answer**   84 ft. head, 483 gpm

$$\frac{\text{New diam.}}{\text{Old diam.}} \times Q = \text{New Q}$$

$$\frac{7.25 \text{ in.}}{7.5 \text{ in.}} \times 500 \text{ gpm} = 483 \text{ gpm}$$

$$\left(\frac{\text{New diam.}}{\text{Old diam.}}\right)^2 \times \text{head} = \text{New Head}$$

$$\left(\frac{7.25 \text{ in.}}{7.5 \text{ in.}}\right)^2 \times 90 \text{ ft.} = 84 \text{ ft.}$$

---

4. **Answer**   150 gpm, 19 ft. head, 17 hp change

Refer to curve for head and hp with current impeller; then calculate change.

$$Q = 1800 \text{ gpm}$$

$$TDH = 110 \text{ ft.}$$

$$Bhp = 62 \text{ hp}$$

$$\frac{\text{New diam.}}{\text{Old diam.}} \times Q = \text{New Q}$$

$$\frac{6.5}{6} \times 1800 = \text{New Q}$$

$$1950 \text{ gpm} = \text{New Q}$$

Pump delivers 150 more gallons/minute with larger impeller.

$$\left(\frac{\text{New diam.}}{\text{Old diam.}}\right)^2 \text{ x head } = \text{ New Head}$$

$$\left(\frac{6.5}{6}\right)^2 \text{ x } 110 = \text{ New Head}$$

129 ft. = New Head

Pump lifts water 19 ft. higher with larger impeller.

$$\left(\frac{\text{New diam.}}{\text{Old diam.}}\right)^3 \text{ x Bhp } = \text{ New Bhp}$$

$$\left(\frac{6.5}{6}\right)^3 \text{ x } 62 = \text{ New Bhp}$$

79 hp = New Bhp

Pump uses 17 hp more with larger impeller.

---

5. **Answer**   2.7 psi

Draw diagram. Suction gage reading equals static suction head minus head losses.

Solve for HL:

400 gpm = .89 cfs

Assume C=100

$Q = .435 \ C \ d^{2.63} \ s^{.54}$

$.89 = .435 \ \ 100 \ \ .5^{2.63} \ \ s^{.54}$

$.022 = s$

.022 ft./ft. x 35 ft. = .8 ft. HL

Gage reading = 7 ft. suction head - .8 ft. HL = 6.2 ft. = 2.7 psi

---

6. **Answer**   6 ft. Poor operation; cavitation likely

Solve for Velocity Head.

$Q = 300 \text{ gpm } = .67 \text{ cfs}$

$Q = A \ V$

$.67 = .785 \text{ x } .33^2 \ V$

7.8 ft./sec. = V

$Vh = \dfrac{V^2}{2g} = \dfrac{7.8^2}{64.4} = .95$

$$\text{NPSHA} = (\text{Atmospheric Press.} + \text{Vh}) - (\text{Lift} + \text{Losses} + \text{vapor pressure})$$

$$\text{NPSHA} = (34 + .95) - (15 + 12 + 2)$$

$$\text{NPSHA} = 6 \text{ ft.}$$

$$\text{NPSHA} = \text{NPSHR}$$

---

7. **Answer**   35 ft.

All values are recorded in feet of water Absolute.

We are only considering the suction side of the pump; pump is immersed to 8 ft. in the water (8 ft. suction head); there is no suction lift; suction head must be entered as a negative value.

Vapor pressure: referring to chart, VP at 90° F = 1.6 ft.

Atmospheric Pressure: referring to chart, at 4000 ft. = 29.4 ft.

Velocity Head: solve for velocity (3000 gpm = 6.7 cfs)

$$Q = A \quad V$$

$$6.7 = .785 \times 1^2 \quad V$$

$$8.5 \text{ ft./sec.} = V$$

$$\text{Vh} = \frac{V^2}{2g} = \frac{8.5^2}{64.4} = 1.1 \text{ ft.}$$

$$\text{NPSHA} = (\text{Atmos. Press.} + \text{Vh}) - (\text{Lift} + \text{Losses} + \text{vapor press.})$$

$$\text{NPSHA} = (29.4 + 1.1) - (-8 + 2 + 1.6)$$

$$\text{NPSHA} = 35 \text{ ft.}$$

The pump should operate efficiently as long as NPSHR is less than 35 ft.

---

8. **Answer**   A. 15 ft.
                         B. yes

A. Lift = 15 ft.; Atmos. Press. = 34 ft.; Vapor Pressure = .5 ft.

Solve for Vh, then losses, then NPSHA (800 gpm = 1.8 cfs).

$$Q = A \quad V$$

$$5.1 = .785 \times .67^2 \quad V$$

$$1.8 \text{ ft./sec.} = V$$

$$\text{Vh} = \frac{V^2}{2g} = \frac{5.1^2}{64.4} = .4 \text{ ft.}$$

Assume $C = 100$

$Q = .435 \ C \ d^{2.63} \ s^{.54}$

$1.8 = .435 \ \ 100 \ \ .67^{2.63} \ s^{.54}$

$.02 = s$

$.02$ ft./ft. x 200 ft. $= 4$ ft. HL in suction pipe

NPSHA = (Atmos. Press. + Vh) - (Lift + Losses + vapor press.)

NPSHA = $(34 + .4) - (15 + 4 + .5)$

NPSHA = 15 ft.

B. Since NPSHA (15 ft.) is greater than NPSHR (7 ft.), the pump is adequate to lift the water without cavitation.

NOTE: That when pumping cold water at sea level, the lift and losses are the major factors which decrease the available pressure.

9. **Answer**   A. 192.6 ft.
                 B. 3.3 hp
                 C. 24.9 ft.

Draw diagram.

A. TDH = Lift + Losses

Lift:
30 ft. + pressure at discharge end (56 psi)

56 psi = 129.4 ft.

30 ft. + 129.4 ft. = 159.4 ft.

Losses:
New steel pipe $C = 140$

55 gpm = .123 cfs

$Q = .435 \ C \ d^{2.63} \ s^{.54}$

$.123 = .435 \ \ 140 \ \ .167^{2.63} \ s^{.54}$

$.062 = s$

Minor Loss: 450 ft. discharge pipe + 10 ft. suction pipe

straight pipe   460 ft.
1 open globe   50 ft.
4 elbows   20 ft.
1 borda entrance   + 6 ft.
                    536 ft. of total pipe

.062 ft./ft. x 536 ft. pipe = 33.2 ft. HL in total line

TDH = 30 ft. lift + 33.2 ft. losses + 129.4 ft. pressure = 192.6 ft. TDH

B. Solve for Bhp.

$$\frac{gpm \times head}{3960} \times whp$$

$$\frac{55 \times 192.6}{3960} \times whp$$

$$2.7 = whp$$

$$\frac{2.7\ whp}{.83} = 3.3\ Bhp$$

C. Vapor pressure at 110° F (from chart) = 3 ft.

Solve for Vh:
Q = A V

.123 = .785 x .167² V

5.6 ft./sec. = V

$$Vh = \frac{V^2}{2g} = \frac{5.6^2}{64.4} = .5\ ft.$$

Solve for suction losses:
.062 ft./ft. (slope) x 10 ft. (length suction) = .62 ft. HL in suction

NPSHA = (Atmos. Press. + Vh) - (Lift + Losses + Vapor Press.)

NPSHA = (34 + .5) - (6 + .62 + 3)

NPSHA = 24.9 ft.

---

10. **Answer**   1270 ft.

Draw diagram — Maximum allowable drop in pressure is 5 ft. for suction head plus 6 psi (14 ft.) below atmospheric pressure. (19 ft. drop allowed). Lift and losses must not exceed this.

Lift: 1 ft. for every 100 ft.

Losses: 5 ft. for fitting + 1 ft. for every 1000 ft. of length.

First, subtract fitting loss from maximum allowable pressure drop. This loss will be present no matter how long the pipe is.

19 ft. - 5 ft. = 14 ft.

The lift plus friction losses must not exceed 14 ft.

First try:

If 1400 ft. of pipe will lift the water 14 ft., then this length alone, with no friction losses, will drop the pressure 14 ft.

Friction losses in this length of pipe would be 1.4 ft. (1 ft./1000 ft.). This yields 15.4 ft. of pressure loss. No good.

Second try:

1250 ft. of pipe would lift the water 12.5 ft. This drops the pressure 12.5 ft. Friction loss in 1250 ft. of pipe is 1.25 ft., yielding a total of 13.75 ft. pressure loss. We can do better.

Third try:

1270 ft. of pipe lifts the water 12.7 ft., dropping pressure 12.7 ft. Friction loss in 1270 ft. pipe is 1.27 ft. This yields a pressure drop of 13.97 ft. Close enough.

---

11. **Answer**   73%

Draw diagram. With metered flow, refer to curve for TDH. Calculate TDH; should be the same as on curve. Refer to curve for Bhp & pump efficiency at that flow. Calculate Mhp; calculate motor efficiency, then wire to water efficiency.

Flow = 1600 gpm (3.6 cfs)

From curve, TDH = 125 ft.

Bhp = 59 hp

pump efficiency = 85%

Calculate TDH:
$$Q = .435 \quad C \quad d^{2.63} \quad s^{.54}$$
$$3.6 = .435 \quad 110 \quad .83^{2.63} \quad s^{.54}$$
$$.02 = s$$

.02 ft./ft. x 2500 ft. = 50 ft. HL due to friction

TDH = Lift + Losses

TDH = 75 + 50

TDH = 125 ft.

OK. This checks out.

Calculate Mhp:
.85 Kw/min. x 60 min./hr. = 51 Kw/hr.

$\dfrac{51 \text{ Kw/hr.}}{.746 \text{ Kw./hp}}$ = 68.4 Mhp

Motor efficiency = $\dfrac{\text{Bhp}}{\text{Mhp}}$ = $\dfrac{59}{68.4}$ = 86% efficient

Motor eff. x pump eff. = wire to water efficiency

.86 x .85 = .73

Field efficiency = 73%

chapter 14

1. **Answer**   8.81 gallons

Volume of the chamber is moved with each stroke; solve for volume.

Vol. = .785 d^2 h

Vol. = .785 1^2 1.5

Vol. = 1.178 cu.ft.

Vol. = 8.81 gal.

2. **Answer**   349633 gpd

Solve for volume; change to pumpage.

Vol. = .785 d^2 h

Vol. = .785 $.83^2$ 1

Vol. = .541 cu.ft.

.541 cu.ft. x 1 stroke/sec. x 60 sec./min. x 1440 min./day = 46742.4
cu.ft./day = 349633 gpd

3. **Answer**   21086 gpd

Find vol. chamber; mult. by % stroke; change to pumpage.

Vol. = .785   d² h

Vol. = .785   1²   1.33

Vol. = 1.044 cu.ft.

Vol. = 7.81 gal.

7.81 gal. x .75 = 5.86 gal.

5.86 gal./str. x 1 str./sec. x 60 sec./min. x 15 min./cycle x 4 cycles/day = 21086 gal./day

---

4. **Answer**   4.8 ft.

Solve for vol. chamber; get pumpage/2 mo.; solve for liquid vol. in dig.

Vol. = .785   d² h

Vol. = .785   .5² x .67

Vol. = .1315 cu.ft.

.1315 cu.ft./str. x .5 str./sec. x 60 sec./min. x 10 min./cycle x 6 cycles/day x 60 days = 9468 cu.ft.

Vol. = .785   d² h

9468 = .785   50² h

4.8 ft. = h

---

5. **Answer**   A. 12.8 gpm
B. 92,312 gal./mo.

Solve for vol. chamber; mult. by RPM.

A. Vol. = .785   d² h

Vol. = .785   .432²   .18

Vol. = .0264 cu.ft.

Vol. = .197 gal.

.197 gal./str. x 65 str./min. = 12.8 gpm

B.   12.8 gpm x 60 min./hr. x 4 hr./day x 30 day./mo. = 92,312 gal./mo.

6. **Answer**   A. 9.4 ft.
                 B. Filling - by .18 cfs

A. Solve for volume chamber; get pumpage/4 hr.; use vol. tank to solve for h.

Vol. = .785   d²   h

Vol. = .785   .5²   .5

Vol. = .098 cu.ft.

.098 cu.ft./str. x 500 str./min. x 60 min./hr. x 4 hr. = 11775 cu.ft./4 hr.

Vol. = .785   d²   h

11775 = .785   40²   h

9.4 ft. = h = drawdown

B. Solve for flow (cu.ft. pumped/sec.) out; compare against flow in.

$\dfrac{500 \text{ str./min.}}{60 \text{ sec./min.}}$ = 8.33 str./sec.

.098 cu.ft./str. x 8.33 str./sec. = .82 cu.ft./sec.

1 cfs in. - .82 cfs out = .18 cfs in

Tank is filling - by .18 cfs.

---

7. **Answer**   A. 85% capacity
                 B. checks slipping; piston worn; set at full stroke?

A. Solve for pumped volume at 100% capacity; solve for volume actually pumped (drawdown); compare.

Vol. = .785   d²   h

Vol. = .785   .67²   .67

Vol. = .236 cu.ft.

.236 cu.ft./stroke x 50 strokes/min. x 7.48 gal./cu.ft. x 45 min. x 2 pumps = 7947 gallons pumped theoretical

Vol. = .785   d²   h

Vol. = .785   24²   2

Vol. = 904.3 cu.ft.

Vol. = 6764.3 gallons drawdown

6764.3 gal. drawdown/7947 theoretical pumpage = 85% capacity

APPENDIX II

CONVERSIONS

Weight to Volume: 1 gram = 1 ml. (water only)

8.34 lb. = 1 gallon (water only)

62.4 lb. = 1 cubic foot (water only)

Volume: 7.48 gal. = 1 cu.ft.

1000 liters = 1 cubic meter

3.785 liters = 1 gallon

Linear: 5280 ft. = 1 mile

Time: 1440 minutes = 1 day

Flow: 1 MGD = 1.55 cfs

Pressure: 1 psi = 2.31 ft. of water depth

1 ft. of water depth = .433 psi

Power: 1 hp = .746 Kw

APPENDIX III

IMPORTANT FORMULAS

Area of a circle: $A = .785 \ d^2$

Perimeter of a circle: $P = 3.14 \ d$

Volume of a cylinder: $Vol. = .785 \ d^2 \ h$

Surface Area of Cylinder Wall: $A = 3.14 \ d \ h$

Pressure: $P = w \ h$

Force: $F = P \ A$

Equation of continuity: $Q = A \ V$

Bernoulli's Formula: $Ph_1 + Vh_1 + Z_1 = Ph_2 + Vh_2 + Z_2 + HL$

Velocity Head: $Vh = \dfrac{V^2}{2g}$

Horsepower: $\dfrac{gpm \ x \ hd}{3960} = whp$

Hazen-Williams Formula: $Q = .435 \ C \ d^{2.63} \ s^{.54}$

Slope: $s = \dfrac{HL}{L}$

Manning's Formula: $V = \dfrac{1.486}{n} \ R^{.66} \ s^{.5}$

Hydraulic Radius: $R = \dfrac{wetted \ area}{wetted \ perimeter}$

Hydraulic Radius (full pipe): $R = \dfrac{d}{4}$

Flow through orifice: $Q = C_d \ A \ \sqrt{(Ph_1 - Ph_2) \ x \ 2g}$

Flow through venturi:

$$Q = C_d \ A \ \sqrt{\frac{(Ph_1 - Ph_2) \times 2g}{1 - \left(\frac{d_2}{d_1}\right)^4}}$$

Weir Flow (rectangular): $\quad Q = 3.33 \ L \ h^{1.5}$

Weir flow (90° V-notch): $\quad Q = 2.5 \ h^{2.5}$

Flow through Parshall Flume: $\quad Q = 4 \ W \ H_a^{1.52} \ W^{.026}$

Trajectory of a jet: $\quad Q = \frac{4 \ A \ X}{\sqrt{y}}$

Available Net Pos. Suction Head: $\quad$ NPSHA =
(Atmos. Press. + Vh) - (Lift + Loss + vp)

APPENDIX IV

FORMULA DERIVATIONS

Velocity Head Formula

From basic physics, when force is applied to an object, it will travel a distance, y, equal to half the acceleration of the object multiplied by the square of the time involved.

$$y = \tfrac{1}{2} a t^2 \qquad \text{This distance is the Velocity Head.}$$

The acceleration is the force of gravity, and t can be converted to V/g (Vel. = accel. x time), (V = gt), (t = V/g)

so:
$$y = \tfrac{1}{2} a t^2$$
$$y = \tfrac{1}{2} g \times V^2/g^2$$
$$y = \tfrac{1}{2} V^2/g$$
$$y = V^2/2g$$

Horsepower Formula

Power = Power

ft.lb./min. = ft.lb./min.

Now separate the components of one side of the equation.

feet of lift x lb. of water flow/min. = ft.lb./min.

Divide right side by 33000 ft. lb./min./hp to change it to hp - and divide left side by 8.34 lb./gal. to change lb. of water to gal. of water.

$$\frac{\text{ft. x lb./min.}}{8.34 \text{ lb./gal.}} = \frac{\text{ft.lb./min}}{33000 \text{ ft.lb./min./hp.}}$$

$$\frac{\text{ft. x gal./min.}}{3960} = \text{hp}$$

Hazen-Williams Formula

Take Darcy-Weisbach formula:

$$HL_f = f \times L \times \frac{1}{d} \times \frac{V^2}{2g}$$

Rearrange it; move L to the left.

$$\frac{HL_f}{L} = f \times \frac{1}{d} \times \frac{V^2}{2g}$$

HL/L is slope; move it back to the right side as slope; move V to the left side; invert the formula:

$$V^2 = s \times \frac{d}{f} \times 2g$$

Take the square root of every value:

$$V = \frac{s^{.5}}{f^{.5}} \times d^{.5} \times 2g^{.5}$$

$$V = \frac{Q}{A}$$

Equate V to Q:

$$Q = A \times \frac{s^{.5}}{f^{.5}} \times d^{.5} \times 2g^{.5}$$

Enter numerical values for A & 2g:

$$Q = .785 \ d^2 \times \frac{s^{.5}}{f^{.5}} \times d^{.5} \times 64.4^{.5}$$

Change f to C; condense: $(f = 200/C^2)$

$$Q = .445 \times C \times d^{2.5} \times s^{.5}$$

Orifice Meter Flow Formula

Bernouilli's Formula

$$Ph_1 + Vh_1 + Z_1 = Ph_2 + Vh_2 + Z_2 + HL_f$$

Eliminate Z_1 & Z_2, and consider Vh_1 as negligible

$$Ph_1 = Ph_2 + Vh_2 + HL_f$$

Convert Vh_2 to $V^2/2g$

$$Ph_1 - Ph_2 = \frac{V^2}{2g} + HL_f$$

Equate to velocity

$$\sqrt{(Ph_1 - Ph_2) \times 2g} = V + HL_f$$

Equate to flow

$$A\sqrt{(Ph_1 - Ph_2) \times 2g} = Q + HL_f$$

Coefficient of discharge replaces HL_f

$$Q = C_d A\sqrt{(Ph_1 - Ph_2) \times 2g}$$

Venturi Meter Flow Formula

Bernoulli's Formula:

$$Ph_1 + Vh_1 + Z_1 = Ph_2 + Vh_2 + Z_2 + HL_f$$

Eliminate Z_1 & Z_2

$$Ph_1 + Vh_1 = Ph_2 + Vh_2 + HL_f$$

$$Ph_1 - Ph_2 = Vh_2 - Vh_1 + HL_f$$

Convert to $V^2/2g$

$$Ph_1 - Ph_2 = \frac{V^2_2 - V^2_1}{2g} + HL_f$$

From $AV = AV$, we can derive

$$V_1 = \frac{d_2}{d_1} \times V_2 \quad \text{and} \quad V^2_1 = \frac{d^4_2}{2d_1} \times V^2_2$$

Substituting into formula

$$Ph_1 - Ph_2 = \frac{V^2_2 - \left(\frac{d_2}{d_1}\right)^4 \times V^2_2}{2g} + HL_f$$

Solving for V_2 yields

$$Q = C_d \ A \ \sqrt{\frac{(Ph_1 - Ph_2) \times 2g}{1 - \left(\frac{d_2}{d_1}\right)^4}}$$

Convert to Q, and add C_d

$$Q = C_d \ A \ \sqrt{\frac{(Ph_1 - Ph_2) \times 2g}{1 - \left(\frac{d_2}{d_1}\right)^4}}$$

D_1 = upstream diameter

D_2 = throat diameter

Trajectory of a Jet Formula

From basic physics, the distance of travel of a falling body, y, is equal to half its acceleration multiplied by the square of the time it falls:

$$y = \tfrac{1}{2}\, a\, t^2$$

Acceleration = acceleration due to gravity:

In addition, a body traveling horizontally, will travel a distance, x, equal to its speed times the time it travels:

$$x = \text{Velocity times time} \; - \text{or} - \; x/V = t$$

These two formulas can be combined and simplified in order to create one which includes both x and y, the measurable quantities:

$$y = \tfrac{1}{2}\; g\; \frac{x^2}{V^2}$$

Equate to V:

$$V^2 = \tfrac{1}{2}\; g\; \frac{x^2}{y}$$

Use value of g and combine:

$$V^2 = 16\; \frac{x^2}{y}$$

Eliminate exponents and equate to Q:

$$Q = 4\; A\; \frac{x}{\sqrt{y}}$$

APPENDIX V

NUMERIC TABLES

FLOW CONVERSIONS

gal/min	gal/day	MGD	cfs
139	200,000	.2	.31
278	400,000	.4	.62
417	600,000	.6	.93
556	800,000	.8	1.24
694	1,000,000	1.0	1.55
1042	1,500,000	1.5	3.33
1389	2,000,000	2.0	3.10
1736	2,500,000	2.5	3.88
2083	3,000,000	3.0	4.65
2431	3,500,000	3.5	5.43
2778	4,000,000	4.0	6.20
3125	4,500,000	4.5	6.98
3472	5,000,000	5.0	7.75
3819	5,500,000	5.5	8.53
4167	6,000,000	6.0	9.30
4514	6,500,000	6.5	10.08
4861	7,000,000	7.0	10.85
5208	7,500,000	7.5	11.63
5556	8,000,000	8.0	12.40
5903	8,500,000	8.5	13.18
6250	9,000,000	9.0	13.95
6597	9,500,000	9.5	14.73
6944	10,000,000	10.0	15.50

WATER MAIN FLUSHING - GPM

10 ft/sec.	391	612	881	1567	2427	3523	4799	6260	7927
8 ft/sec.	313	489	705	1254	1942	2818	3838	5008	6342
5 ft/sec.	195	306	441	784	1214	1762	2399	3130	3964
3 ft/sec.	117	184	264	470	728	1057	1439	1878	2378
2 ft/sec.	78	122	176	314	485	705	960	1252	1585
	4	5	6	8	10	12	14	16	18

MAIN DIAMETER (INCHES)

MAXIMUM FIRE FLOW (GPM)

STATIC PRESSURE												
70 psi	564	1128	1692	606	1212	1819	659	1318	1977	727	1454	2181
65 psi	573	1145	1718	622	1245	1867	687	1374	2060	775	1550	2324
60 psi	584	1168	1752	645	1289	1933	727	1454	2181	849	1698	2548
55 psi	599	1199	1799	676	1353	2029	790	1580	2370	984	1967	2951
50 psi	622	1245	1876	727	1454	2181	905	1810	2714	---	---	---
45 psi	659	1318	1977	820	1640	2460	---	---	---	---	---	---
40 psi	727	1454	2181	---	---	---	---	---	---	---	---	---
Q, gpm	500	1000	1500	500	1000	1500	500	1000	1500	500	1000	1500
	30 psi			35 psi			40 psi			45 psi		

TEST PRESSURE

(Table calculation based on fire flow pressure drop to 20 psi)

WASTEWATER WET WELL DRAWDOWN (GPM)

(chart can be used to calculate volume change in a tank for any reason)

1. Shut off pumps; measure rise rate (in./min.).
2. Refer to chart for gpm rise.
3. Turn on pumps; measure drop rate (in./min.).
4. Refer to chart for gpm drop.
5. RISE RATE(gpm) + DROP RATE(gpm) = PUMPING RATE(gpm)

IN./MIN. RISE/DROP

RISE/DROP	25	30	35	40	45	50	55	60	65	70	75	80	85	90
8	2447	3523	4796	6264	7927	9787	11842	14093	16540	19182	22020	25054	28284	31709
7	2141	3083	4196	5481	6936	8563	10362	12331	14472	16784	19268	21922	24748	27745
6	1835	2642	3597	4698	5945	7340	8881	10570	12405	14386	16515	18790	21213	23782
5	1529	2202	2997	3915	4954	6117	7401	8808	10337	11989	13763	15659	17677	19818
4	1223	1762	2398	3132	3964	4893	5921	7046	8270	9591	11010	12527	14142	15854
3	917	1321	1798	2349	2973	3670	4441	5285	6202	7193	8258	9395	10606	11891
2	612	881	1199	1566	1982	2447	2961	3523	4135	4796	5505	6264	7072	7929
1	306	440	599	783	991	1223	1480	1762	2067	2398	2752	3132	3535	3963
	25	30	35	40	45	50	55	60	65	70	75	80	85	90

WET WELL DIAMETER (FT.)

POWER CALCULATIONS

1. Use pump curve to obtain brake horsepower.
2. Look up Motor Horsepower on Chart #1.
3. Obtain cost per unit time on Chart #2.

CHART #1 - MOTOR HORSEPOWER

BHP

10.0	10.5	11.1	11.8	12.5	13.3	14.3	15.4
9.5	10.0	10.5	11.2	11.8	12.7	13.5	14.6
9.0	9.5	10.0	10.6	11.3	12.0	12.9	13.9
8.5	9.0	9.4	10.0	10.6	11.3	12.1	13.1
8.0	8.4	8.9	9.4	10.0	10.7	11.4	12.3
7.5	7.9	8.3	8.8	9.4	10.0	10.7	11.5
7.0	7.4	7.8	8.2	8.8	9.3	10.0	10.8
6.5	6.8	7.2	7.7	8.1	8.7	9.3	10.0
6.0	6.3	6.7	7.1	7.5	8.0	8.6	9.2
5.5	5.8	6.1	6.5	6.9	7.3	7.9	8.5
5.0	5.3	5.6	5.9	6.3	6.7	7.1	7.7
4.5	4.7	5.0	5.3	5.6	6.0	6.4	6.9
4.0	4.2	4.4	4.7	5.0	5.3	5.7	6.2
3.5	3.7	3.9	4.1	4.4	4.7	5.0	5.4
3.0	3.2	3.3	3.5	3.8	4.0	4.3	4.6
2.5	2.6	2.8	2.9	3.1	3.3	3.6	3.9
2.0	2.1	2.2	2.4	2.5	2.7	2.9	3.1
1.5	1.6	1.7	1.8	1.9	2.0	2.1	2.3
1.0	1.1	1.1	1.2	1.3	1.3	1.4	1.5
.5	.5	.6	.6	.6	.7	.7	.8
	.95	.90	.85	.80	.75	.70	.65

% MOTOR EFFICIENCY

CHART #2 - COST PER UNIT TIME
FOR EVERY $.01/KW HR

MHP

	DAY	WEEK	MONTH	YEAR
15.0	2.69	18.80	80.57	980.24
14.0	2.51	17.54	75.20	914.89
13.0	2.33	16.29	69.83	849.55
12.0	2.15	15.04	64.45	784.20
11.0	1.97	13.78	59.08	718.84
10.0	1.79	12.53	53.71	653.50
9.0	1.61	11.28	48.34	588.15
8.0	1.43	10.02	42.97	522.80
7.0	1.25	8.77	37.60	457.45
6.0	1.07	7.52	32.22	392.10
5.0	.90	6.27	26.86	326.75
4.5	.81	5.64	24.17	294.08
4.0	.72	5.01	21.49	261.40
3.5	.63	4.39	18.80	228.73
3.0	.54	3.76	16.11	196.05
2.5	.45	3.13	13.43	163.38
2.0	.36	2.51	10.74	130.70
1.5	.27	1.88	8.06	98.03
1.0	.18	1.25	5.37	65.35
0.5	.09	.63	2.69	32.68
	DAY	WEEK	MONTH	YEAR

FRICTION LOSS (FT.) PER MILE OF PIPE

gpm	4 inches			6 inches			8 inches			10 inches			12 inches			14 inches			16 inches			18 inches		
C=	120	100	80	120	100	80	120	100	80	120	100	80	120	100	80	120	100	80	120	100	80	120	100	80
900	2662	3749	5596	379	534	790	88	124	185	31	44	66	13	18	27	6	9	13	3	4	6	2	3	5
850	2405	3379	5043	336	473	706	82	116	173	28	39	58	11	16	24	5	7	11	3	4	6	1	2	3
800	2137	3010	4492	298	420	627	71	100	154	25	35	52	10	14	21	5	7	11	2	3	5	1	2	3
750	1912	2693	4019	263	376	561	65	92	137	22	31	46	9	13	19	4	6	9	2	3	5	1	2	3
700	1683	2371	3539	232	327	488	57	80	121	20	28	42	9	12	17	4	5	8	2	3	5	1	2	3
650	1462	2059	3073	202	285	425	48	68	102	17	24	36	7	10	15	4	5	8	1	2	3	.7	1	2
600	1263	1779	2651	175	246	367	43	60	89	15	21	31	6	9	13	3	4	6	1	2	3	.7	1	2
550	1076	1515	2257	148	209	311	36	51	76	12	17	25	5	7	10	3	4	6	1	2	3	.7	1	2
500	900	1267	1888	124	174	264	31	43	64	11	15	22	4	6	9	2	3	5	1	2	3	.7	1	2
450	742	1045	1557	103	145	216	26	36	54	9	12	18	4	5	8	1	2	3	.7	1	2	.6	.8	1
400	600	845	1256	82	116	173	21	29	43	7	10	15	3	4	6	1	2	3	.7	1	2	.4	.6	.9
350	465	655	986	71	90	136	16	23	34	5	7	11	2	3	5	1	2	3	.6	.8	1.2	.4	.5	.8
300	346	487	726	50	70	104	12	17	25	4	6	8	1	2	3	.7	1	2	.4	.6	.9	.2	.3	.5
250	251	354	528	36	50	73	9	12	18	3	4	6	1	2	3	.6	.8	1.2	.3	.4	.6	.1	.2	.3
200	164	232	346	23	33	49	6	8	12	2	3	5	.7	1	1	.4	.5	.8	.2	.3	.5	.1	.1	.3
150	98	138	206	14	19	28	4	5	8	1	2	3	.4	.6	.9	.2	.3	.5	.1	.2	.3	.1	.1	.1
100	45	64	95	6	9	13	1	2	3	.6	.8	1	.2	.3	.5	.07	.1	.2	.1	.1	.2	.03	.04	.1
50	4	6	9	1	2	3	.4	.6	.9	.1	.2	.3	.06	.08	.1	.03	.04	.06	.01	.02	.03	.01	.01	.0

FEET OF PIPE EQUIVALENT TO FITTINGS

Fitting	2	3	4	5	6	8	10	12	14	16	18	20	22	24	30	36
globe valve	55	85	105	140	160	220	280	330	390	440	500	550	600	650	850	1000
angle valve	27	40	55	70	90	120	140	160	190	220	250	280	300	340	420	500
gate valve:open	1.2	1.7	2.3	3	4	5	6	7	8	9	11	12	13	15	18	20
gate valve: 1/2closed	35	50	70	90	110	130	160	190	220	260	300	330	360	400	500	580
gate valve: 3/4closed	150	200	250	350	400	550	700	800	1050	1200	1300	1400	1500	1700	2000	2500
swing check valve	13	20	26	32	40	50	65	75	90	105	120	130	140	160	190	240
close return bend	13	18	24	30	36	48	60	72	85	100	112	125	135	150	180	220
std.tee:side outlet	12	17	22	28	33	38	50	65	75	85	100	110	125	140	160	200
std.tee: run	3.3	5	7	9	11	14	17	20	24	28	31	35	38	42	54	60
std. elbow	5.5	8	11	14	17	22	28	33	38	43	48	50	54	65	80	100
long sweep elbow	3.3	5	7	9	11	14	17	20	24	28	31	35	38	42	54	60
square elbow	12	17	22	28	33	38	55	65	75	85	100	110	125	140	160	200
45 degree elbow	2.5	3.8	5	6	7.5	10	13	16	18	20	23	25	28	31	38	45
Borda entrance	5	7.5	10	13	15	19	25	29	34	40	43	48	54	60	75	90
ordinary entrance	3	4	6	7.5	9	12	15	18	20	22	25	28	31	35	43	50
ordinary exit	5	7.5	10	13	15	19	25	29	34	40	43	48	54	60	75	90
sudden enlarg. 1:4	5.5	8	11	14	17	22	28	33	38	43	48	50	54	65	80	100
sudden enlarg. 1:2	3	4	6	8	10	13	16	18	22	25	28	32	35	39	48	58
sudden enlarg. 3:4	1.2	1.7	2.3	3	4	4.5	6	7	8	9	11	12	13	15	18	20
sudden contract. 4:1	2.5	3.8	5	6	7.5	10	13	16	18	20	23	25	28	31	38	45
sudden contract. 2:1	1.8	3	4	5	6	8	10	12	14	15	16	18	20	24	28	34
sudden contract. 4:3	1.2	1.7	2.3	3	4	4.5	6	7	8	9.5	11	12	13	15	18	20

PIPE DIAMETER (INCHES)

FLOW THROUGH VENTURI METER (CFS)

DIFF.PRESS	6" diam.pipe				8" diam.pipe				10" diam.pipe			12" diam.pipe			18" diam. pipe		
8 psi	.7	1.7	3.3	6.4	1.7	3.0	5.0	8.0	3.0	7.1	15.5	6.9	13.2	26.7	29.6	45.4	76.9
7 psi	.7	1.6	3.1	6.0	1.6	2.8	4.7	7.5	2.8	6.7	14.5	6.4	12.3	23.6	27.7	42.5	72.0
6 psi	.6	1.5	2.8	5.6	1.5	2.6	4.3	6.9	2.6	6.2	13.4	5.9	11.4	21.8	25.7	39.3	66.6
5 psi	.6	1.4	2.6	5.1	1.3	2.4	4.0	6.3	2.4	5.6	12.2	5.4	10.4	19.9	23.4	35.9	60.8
4 psi	.5	1.2	2.3	4.5	1.2	2.1	3.5	5.7	2.1	5.0	10.9	4.8	9.3	17.8	21.0	32.1	54.4
3 psi	.5	1.1	2.0	3.9	1.0	1.9	3.1	4.9	1.8	4.4	9.5	4.2	8.1	15.5	18.1	27.8	47.1
2 psi	.4	.9	1.6	3.2	.8	1.5	2.5	4.0	1.5	3.6	7.7	3.4	6.6	12.6	14.8	22.7	38.4
1 psi	.3	.6	1.2	2.3	.6	1.1	1.8	2.8	1.1	2.5	5.5	2.4	4.7	8.9	10.5	16.0	27.2
	2	3	4	5	3	4	5	6	4	6	8	6	8	10	12	14	16

THROAT DIAMETER (INCHES)

FLOW THROUGH ORIFICE METER (CFS)

DIFFERENTIAL PRESS.(PSI)	.5"		1"		2"		3"		4"		5"		6"		7"		8"		10"	
$C_d =$	.6	.8	.6	.8	.6	.8	.6	.8	.6	.8	.6	.8	.6	.8	.6	.8	.6	.8	.6	.8
10	.03	.04	.13	.17	.51	.68	1.1	1.5	2.0	2.7	3.2	4.2	4.5	6.1	6.2	8.3	8.1	10.8	12.5	16.7
9	.03	.04	.12	.16	.48	.64	1.1	1.4	1.9	2.6	3.0	4.0	4.3	5.8	5.9	7.8	7.7	10.2	11.9	15.8
8	.03	.04	.11	.15	.45	.61	1.0	1.4	1.8	2.4	2.8	3.8	4.1	5.4	5.5	7.4	7.2	9.6	11.2	14.9
7	.03	.04	.11	.14	.42	.57	.95	1.3	1.7	2.3	2.6	3.5	3.8	5.1	5.2	6.9	6.8	9.0	10.5	14.0
6	.03	.03	.10	.13	.39	.52	.88	1.2	1.6	2.1	2.5	3.3	3.5	4.7	4.8	6.4	6.3	8.4	9.7	12.9
5	.02	.03	.09	.12	.36	.48	.80	1.1	1.4	1.9	2.2	3.0	3.2	4.3	4.4	5.8	5.7	7.6	8.9	11.8
4	.02	.03	.08	.11	.32	.43	.72	.96	1.3	1.7	2.0	2.7	2.9	3.8	3.9	5.2	5.1	6.8	7.9	10.6
3	.02	.02	.07	.09	.28	.37	.62	.83	1.1	1.5	1.7	2.3	2.5	3.3	3.4	4.5	4.4	5.9	6.9	9.2
2	.01	.02	.06	.08	.23	.30	.51	.68	.90	1.2	1.4	1.9	2.0	2.7	2.8	3.7	3.6	4.8	5.6	7.5
1	.01	.01	.04	.05	.12	.21	.36	.48	.64	.85	1.0	1.3	1.4	1.9	2.0	2.6	2.6	3.4	3.3	5.3

ORIFICE DIAMETER

PUMP TROUBLESHOOTING

No Water Delivered

pump not primed - or incompletely primed

discharge head too high - running over shutoff head - check friction losses

suction lift too high - check gage and size of suction pipe

inlet of suction pipe insufficiently submerged

air in through stuffing box or suction piping

plugged pump or piping

rotation or impeller backwards

Not Enough Water Delivered

insufficient prime

partial plugging

speed too slow

discharge or suction lift too high

air in through packing or suction piping (check packing with match; replace packing; check position of lantern ring; water seal piping may be plugged)

foot valve too small or plugged - check strainer

entrained air coming in through suction tank

drawdown too great (well pump)

defective or worn impeller or wear rings

Discharge Pressure Too Low

entrained air

speed too low

impeller diameter too small

impeller clearance too great

discharge gage in the wrong place

mechanical defects - deformed casing

Discharge Pressure Too High

too much friction loss - are valves wide open?

Suction Lift Too High

too much friction loss on suction side

Head Increases When Flow Increases

impeller in backwards or pump turning in wrong direction

Pump Works For a While - Then Quits

incomplete priming
air leaks in stuffing box or suction piping
suction lift too high
entrained air
stuffing box defects - water seal piping plugged

Pump Draws Too Much Horsepower

speed too high
mechanical problems - bent or misaligned shaft; tight packing; bad wear rings; faulty
  bearings; deformed casing
head lower than rated - pumping too much liquid
siphoning - if pump discharge is lower than pump, the unit may be pumping and also
  siphoning, yielding a greater discharge than intended
specific gravity or viscosity of liquid other than that which pump was designed for

Motor Speed Too Low

is motor receiving full voltage?

Pump Vibrates - Excessive Noise

air in pump - any reason
clogging - pump; piping; foot valve
shaft misaligned - or bent
foundations not rigid
worn bearings
impeller worn or damaged
rotating part rubbing on stationary part

Pump Overheats & Seizes

no prime
air in pump
operating at too low capacity
shaft misaligned - bearings worn - parts rubbing

GLOSSARY

Absolute Pressure - the entire pressure on a unit area, including atmospheric pressure; recorded as psia.

Air Lift Pump - air feeds into a vertical water discharge pipe, causing the water to rise.

Atmospheric Pressure - the pressure of the atmosphere upon the earth; (14.7 psi).

Axial Pump - water enters and exits the pump on the same plane.

Bernoulli's Theorem - neglecting friction losses, from point to point in a fluid system employing steady state flow, the total energy is the same at every point in the path of flow.

Blower - a pump which pumps air; may be positive displacement or centrifugal.

Bourdon Tube - a pressure gage; pressure source is connected to a spring assembly which registers on a scale.

Brake Horsepower - the power applied to the pump.

Buoyancy - the upward force on a submerged object equal to the weight of the volume of liquid it has displaced.

C Factor - a numerical designation of interior pipe roughness.

Cavitation - constant formation and implosion of vapor cavities in a moving liquid, causing damage to pipes, pumps and fittings.

Centrifugal Pump - a pumping mechanism whose rapidly spinning impeller imparts a high velocity to the water which enters, then converts that velocity to pressure upon exit.

Coefficient of Discharge - (C_d) a multiplication factor which equates to friction loss and which is used in calculating flow through an orifice; the value of C_d depends on shape of the orifice.

Compound Meter - a meter which includes both a positive displacement meter to measure small flows, and a turbine meter to measure large flows.

Compound Pipe Systems - an arrangement of two or more different pipes, either connected end to end, or side by side.

Compressibility - a reduction in volume of fluid which occurs with the application of external force, such as pressure. Gases are easily compressible; most liquids are not. For all practical purposes, water is an incompressible fluid. With an application of 100 psi pressure, water occupies a volume of only .0000023 cubic feet less than it does at atmospheric pressure. This is very important in the design of piping systems which must handle stresses which the water cannot absorb by compression.

Density - amount of mass per unit volume of a substance.

Design Point - the conditions of head and flow at which a particular centrifugal pump is most efficient.

Diaphragm Pump - positive displacement pump of the plunger type which employs a flexible diaphragm as the pumping mechanism.

Double Suction Pump - water entering approaches the impeller from both sides.

Drawdown - the distance the water level drops once pumping begins.

Dynamic Discharge Head - the vertical distance between the pump centerline and the water level on the discharge side, plus the friction head loss.

Dynamic Suction Head - the vertical distance between the water level on the suction side and the centerline of the pump below it, minus the friction head loss.

Dynamic Suction Lift - the vertical distance between the water level and the centerline of the pump above it, plus the friction head loss.

Efficiency - the power produced by the pump, divided by the power used in operating the pump.

Elevation Head - pressure due to the elevation of the water.

Energy - there are two kinds of energy, kinetic and potential. In a water system, kinetic energy is demonstrated by the velocity of the fluid. Potential energy is present as elevation of the fluid above a reference point, or as pressure. The summation of these three possible energies in a water system is expressed in an equation known as <u>Bernoulli's Theorem</u>, which is the basis for all hydraulic calculations.

Energy Grade Line - level of system energy; it includes Pressure Head and Velocity Head.

Equation of Continuity - (Q=AV); at a given flow, water velocity is dependent upon the cross sectional area of flow.

Equivalent Pipe Theory - two pipes are equivalent if the head loss generated by the water velocity is the same in both pipes.

Field Efficiency - wire to water efficiency; motor efficiency multiplied by pump efficiency.

Flow - the quantity of fluid which passes a point in a given unit of time. Flow can be viewed as a moving volume.

Flow Nets, Flow Models - these are complex patterns simulating the changes in water pressures and flow within a specified area.

Flow Rate Controller - venturi-valve arrangement which automatically controls flow through a constant rate sand filter.

Fluid - matter which will deform without external stress, and takes the shape of its container. Both liquids and gases are fluids. A fluid is infinitely flexible; it can move rapidly in one place and slowly in another. It can transmit a force in any or all directions.

Foot Valve - a check valve at the end of the suction line which shuts and holds the pump's prime.

Force - pressure upon an entire area; registered in pounds.

Friction - contact of fluid molecules with conduit surfaces, resulting in disorganization and loss of pressure.

Gage Pressure - the pressure above or below atmospheric pressure; recorded as psig.

Head - pressure, registered in feet.

Head Loss - pressure loss due to friction, registered in feet.

Horizontal Pump - pump shaft is in a horizontal position.

Hydraulic Grade Line - the line that connects the piezometric surfaces along a pipeline.

Hydraulic Jump - this is an open channel flow phenomenon caused by a sudden decrease in water velocity. The water surface "jumps" to a greater depth with exaggerated turbulence. The great foaming at the the bottom of a dam, and on a smaller scale, shoreline waves, and those created at the wake of a passing boat are examples of hydraulic jump.

Hydraulic Ram - a set of valves which open and close rapidly, making use of water hammer to lift water.

Hydraulic Radius - wetted area divided by wetted perimeter; a measure of the efficiency with which the conduit can transmit flow.

Hydraulics - the study of the principles that govern the behavior of liquids at rest and in motion. It is the study of the mechanics of water and its control by man.

Hydrology - the study of the naturally occurring waters of the earth. It includes their circulation, distribution, and reaction with their environment.

Hydropneumatic Storage - water storage system which makes use of air pressure instead of elevation to create water pressure.

Jet Pump - specialized type of centrifugal pump designed with a recirculating flow and eductor assembly to maximize Velocity Head, providing extra lift.

Laminar Flow - this is ideal fluid flow; water particles move along straight, parallel paths, in layers or streamlines. There is no turbulence in the water, and no friction loss. This is not typical of normal pipe flow because the water velocity is too great, but it does occur in groundwater flow, and an attempt at laminar flow is demonstrated in the use of tube and plate settlers in settling basins.

Magnetic Meter - flowmeter using a magnetic field to induce a voltage which is proportional to water velocity, and convertible to flow.

Major Head Loss - pressure loss due to friction along the length of pipe.

Manometer - a piezometer tube curved into a U shape and filled with an internal fluid of higher specific gravity than the fluid being measured. One or both ends may be connected to a source of pressure.

Mass - the quantity of matter that a substance contains. Matter is divided into two major classes: Solids and Fluids.

Minor Head Loss - pressure loss due to turbulence at bends, valves, fittings and changes in pipe diameter.

Momentum - the mass of the fluid, multiplied by its velocity. Upon encountering bends, turns, obstructions in the conduit, momentum is decreased, but it is converted to a temporary increase in pressure, which is then dissipated as heat. The effect is thrust, surge, or water hammer. These transient pressures can cause serious damage to pipes, pumps and fittings if not provided for.

Motor Horsepower - the power applied to the motor.

Multi-Stage Pump - more than one impeller on the pump shaft.

Net Positive Suction Head - the total amount of energy, in Absolute feet of water, at the pump centerline.

Open Channel Flow - flow in a conduit which has one side in contact with atmosphere. Open channel flow is caused by the downward slope of the channel, which maintains the velocity. Flow in aqueducts, streams, rivers and sewer pipes is open channel flow.

Orifice Meter - flat plate across the line of flow which passes through a small hole in its center; a pressure differential meter.

Parallel Pipe Systems - two or more pipes of different age, material, size or length, laid side by side, with the flow splitting among them.

Parallel Pumps - pumps set side by side and having separate suctions, for the purpose of increasing the pumped flow.

Pascal's Law - the pressure at any one point in a static liquid is exerted with equal intensity in all directions.

Parshall Flume - constricted section of an open channel through which the flow passes; reading is by depth measurement.

Peristaltic Pump - as the fluid passes through flexible tubing, an external moving element progressively flattens the tubing, pushing the liquid forward.

Piezometer - an open-ended vertical tube attached at the point of pressure measurement; this is the simplest form of pressure gage.

Piston Pump - single piston (plunger) set in a chamber with a check valve on either side; it is merely an expansion of the pipe adapted to provide a force against the liquid in one direction.

Pitot Gage - a flow measurement device consisting of a double pressure gage, one having an end pointed upstream. Velocity Head is calculated by measurement of pressure differential.

Pneumatic Ejector - pumping apparatus which makes use of compressed air to expel water from a chamber.

Positive Displacement Meter - measuring chamber encloses a disc or piston which oscillates to pass the flow and transmits the movement to a register, which records.

Power - the rate at which work is done; (ft.lb./sec).

Pressure - an amount of force acting on a unit area.

Pressure Head - pressure due to the depth of water.

Priming Chamber - liquid reservoir above the pump which empties to fill the pumping chamber upon startup.

Progressive Cavity Pump - a positive displacement type in which a rotary motion opens a cavity which moves forward along the pump length.

Propeller Meter - for measuring mainline flow, chamber encloses a small propeller which responds to velocity.

Proportional Meter - bypasses a small percentage of the flow through the meter and measures it; the measurement is then multiplied up to the actual flow.

Pump Affinity Laws - a method of calculating the effect that changes in pumping speed have on head, flow and horsepower requirements.

Pump Bearing - rotating mechanism attached to the pump shaft which absorbs pumping thrust.

Pump Characteristics - the range of heads, flows, horsepowers, and efficiencies at which a particular pump will operate.

Pump Performance Curve - a diagram of how a particular pump will perform over the range of its characteristics.

Pump System - the arrangement of lines and appurtenances through which the liquid will be pumped.

Pumping Water Level - water level when the pump is on.

Radial Pump - flow leaves the pump at a 90 degree angle from the entrance direction.

Radius of Influence - horizontal distance between pump shaft and outermost area of drawdown.

Reducers - fittings at pump suction and discharge which step pipe up or down to pump size.

Rotameter - measures flow based on movement of a disc through a variable area conduit.

Rotary Gear Pump - intermeshing gears trap the liquid and move it from suction to discharge.

Safe Well Yield - maximum pumping rate achievable without increasing the drawdown.

Screw Pump - Archimedes screw set at an incline and used to lift water to a higher elevation.

Series Pipe Systems - two or more pipes of different age, material or size, laid end to end.

Series Pumps - pumps or impellers arranged one after the other, to increase pumping head.

Single Stage Pump - one impeller on the pump shaft.

Single Suction Pump - one suction line enters the pump casing.

Slope - in closed pipe systems, the feet of head loss for each foot of pipe.

Solid - matter which holds its shape and will not deform without external stress.

Specific Capacity - pumping rate per foot of drawdown; (gpm/ft.)

Specific Gravity - density of a substance compared to the density of a standard substance.

Static Discharge Head - the vertical distance from the pump centerline to the water level on the discharge side.

Static Suction Head - the vertical distance from the water surface on the suction side to the pump centerline below it.

Static Suction Lift - the vertical distance from the water surface on the suction side to the pump centerline above it.

Static Water Level - water level in a well when the pump is off.

Steady State Flow - flow is steady if at any one point the velocity of the water does not vary and the pressure is constant.

Stuffing Box - the container for the packing rings which encircle the pump shaft, and which prevents pumped liquid from leaking out of the pump.

Submersible Pump - vertical, close-coupled centrifugal pump designed to work with both motor and pump submerged in the water being pumped.

Surge - slow motion mass tendency of the water to "pile up" as momentum decreases; a pressure increase results.

System Curve - a graph which demonstrates what the Total Dynamic Head will be for a particular system over a range of flows.

Total Dynamic Head - the difference between the dynamic suction head and the dynamic discharge head, or the sum of the dynamic suction lift and the dynamic discharge head.

Total Static Head - the difference between the static suction head and the static discharge head, or the sum of the static suction lift and the static discharge head.

Thrust - transient pressure created at turns in a pipeline; tends to push the fitting loose.

Thrust Block - a mass of concrete cast in place onto the pipe and outward from a turn; its purpose is to stabilize the joint.

Turbine Generator - apparatus which makes use of potentially wasted energy to generate electricty.

Turbine Meter - flow passes through a chamber enclosing a rotor which turns in response to velocity.

Turbulent Flow - turbulent flow is normal for a water system; water particles move in a haphazard fashion and continually cross each other in all directions; pressure losses result along a length of pipe because of it.

Ultra Sonic Meter - uses ultrasound frequency differential to detect water velocity or depth which is then convertible to flow.

Uniform Flow - occurs when the magnitude and direction of velocity do not change from point to point. The flow of water through long pipelines of constant diameter is uniform.

Vacuum Pump - a pump which empties an air chamber, or which creates a vacuum by eductor action.

Vapor Pressure - the partial pressure created by water vapor molecules as the liquid evaporates.

Variable Speed Pump - shaft rotation speed varies with system needs.

Velocity - the speed at which a liquid moves.

Velocity Head - energy present due to the velocity of the water.

Venturi Meter - a section of constricted pipe through which the flow passes; a pressure differential meter.

Vertical Pump - pump shaft is in a vertical position.

Viscosity - internal resistance of the fluid to the motion of its particles. A viscous fluid tends to damp out turbulence; it moves in streamlines and resists the shear force that causes friction. Water is nonviscous, having a viscosity of 1.41×10^{-6} sq.ft./sec.

Water Hammer - temporary increase in pressure resulting from rapid change in velocity or direction of the water.

Water Horsepower - the power necessary to lift water to a required height.

Weir - flat plate across the line of flow in an open channel, and over which the flow passes; reading is by depth measurement.

Wire To Water Efficiency - pump efficiency multiplied by motor efficiency.

SOURCES CONSULTED

TEXTS AND MANUALS

American Water Works Association. <u>Basic Science Concepts and Applications</u>. Denver: American Water Works Association, 1980.

American Water Works Association. <u>Hydraulics and Water Loss Control</u>. Denver: American Water Works Association, 1984.

American Water Works Association. <u>Sizing Water Service Lines and Meters</u>. Denver: American Water Works Association, 1975.

American Water Works Association. <u>Spillway Design and Practice</u>. New York: American Water Works Association, 1966.

American Water Works Association. <u>Water Meters: Selection, Installation, Testing, Maintenance</u>. Denver: American Water Works Association, 1972.

Arasmith, E. <u>Applied Practical Hydraulics</u>. Albany: Linn-Benton Community College, 1978.

Binder, R. <u>Fluid Mechanics</u>. Englewood Cliffs: Prentice-Hall, 1943.

Brankert, K. <u>Elementary Theoretical Fluid Mechanics</u>. New York: John Wiley & Sons, 1964.

Brown, E. <u>Hydraulics for Operators</u>. Stoneham: Butterworth, 1985.

Chaudry, M. <u>Applied Hydraulic Transients</u>. New York: Van Nostrand Reinhold, 1987.

Cheremsinoff, N. <u>Fluid Flow</u>. Ann Arbor: Ann Arbor Science, 1981.

Esposito, A. <u>Fluid Power With Applications</u>. Englewood cliffs: Prentice-Hall, 1980.

French, R. <u>Open Channel Hydraulics</u>. New York: McGraw-Hill, 1985.

Giles, R. <u>Fluid Mechanics and Hydraulics</u>. New York: McGraw-Hill, 1962.

Hammer, J. <u>Hydrology and Quality of Water Resources</u>. New York: John Wiley & Sons, 1981.

Hunsacker, J. and Rightmire, B. <u>Engineering Applications of Fluid Mechanics</u>. New York: McGraw-Hill, 1947.

Karassik, I. <u>Centrifugal Pump Clinic</u>. New York: Marcel Dekker, 1981.

Kanen, J. <u>Applied Hydraulics</u>. New York: Holt, Rinehart & Winston, 1986.

King, H., Wisler, C., and Woodburn, J. <u>Hydraulics</u>. New York: John Wiley & Sons, 1922.

King, H. Handbook of Hydraulics. New York: McGraw-Hill, 1954.

Lambeck, R. Hydraulic Pumps and Motors. New York: Marcel Dekker, 1983.

Lewis Publishers. Submersible Sewage Pumping Systems Handbook. Chelsea: Lewis Publishers, 1985.

Linsley, R. and Franzini, J. Water Resources Engineering. New York: McGraw-Hill, 1964.

McNickle, L. Simplified Hydraulics. New York: McGraw-Hill, 1966.

Parr, B. Treatment Plant Hydraulics for Environmental Engineers. Englewood Cliffs: Prentice-Hall, 1984.

Prasuhn, A. Fundamentals of Hydraulic Engineering. New York: Holt, Rinehart & Winston, 1938.

Smith, P. BASIC Hydraulics. London: Butterworth, 1982.

Stewart, H. Pumps. Boston: T. Audel & Co., 1970.

Valentine, H.R. Water in the Service of Man. Middlesex: Penguin Books, 1967.

Van Haveren, B. Water Resource Measurements. Denver: American Water Works Association, 1986.

Vesilind, P., and Peirce, J. Environmental Engineering. Boston: Butterworth, 1982.

Viessman, W., and Hammer, M. Water Supply and Pollution Control. New York: Harper & Row, 1985.

Walker, R. Pump Selection. Ann Arbor: Ann Arbor Science, 1972.

Walski, T. Analysis of Water Distribution Systems. Berkshire: Van Nostrand Reinhold, 1984.

Water Pollution Control Federation. Design of Wastewater and Stormwater Pumping Stations. Washington DC: Water Pollution Control Federation, 1981.

PERIODICALS

"Accurate Flow Measurements a Necessity". Opflow, February 1985.

"An Engineering Approach To Waterline Thrust Restraint". Public Works, August 1989.

Arasmith, E. Pumps and Pumping. Albany: ACR Publications, 1993.

"Causes and Control of Water Hammer". Opflow, September 1983.

"Construction and Use of a Pitot Gage". Opflow, February 1989.

"Developing System Head Curve for Water Distribution Pumping". Journal, American Water Works Association, July 1989.

Ductile Iron Pipe Research Association. Thrust Restraint Design For Ductile Iron Pipe. Birmingham: Ductile Iron Pipe Research Association, 1992.

French, William. "Understanding and Choosing a Centrifugal Pump". Water Engineering and Management, August 1987.

Gierer, Bill. "Pumps - Mechanical Hearts of the Water Industry". Opflow, April 1982.

Hilsdon, Charles. "Using Pump Curves and Valving Techniques for Efficiency Pumping". Journal, American Water Works Association, March 1982.

Jones, Larry S. "Thrust in Underground Pipe". Water Engineering & Management, November 1988.

"Improved Meter Accuracy Reduces District's Problems". Public Works, July 1989.

Holstrom, John. "Another Look at the Versatile Venturi". Water Engineering & Management, November 1987.

Kroon, J. and Stoner, M. and Junt, W. "Water Hammer: Causes and Effects". Journal, American Water Works Association, November 1984.

"Purposes and Limits of Pressure Reducing Valves". Opflow, January 1988.

Reid, Anthony. "Selecting Pipelines to Achieve Effective Energy Conservation". Water and Sewage Works, November 1980.

Seruga, Ed. "Sizing and Selecting Modern Water Meters". Opflow, July 1986.

Sharp, W. and Walski, T. "Predicting Internal Roughness in Water Mains". Journal, American Water Works Association, November 1988.

Salvina, Louis. "What to Look For in Centrifugal Pump Maintenance". Opflow, February 1987.

Smith, Earl C. "Watch Your Gauges". Operations Forum, August 1986.

Tarrant, J. "Pipe Flow Measurement By Orifice". Water and Sewage Works, 1966.

Walski, T., Gessler, J. and Sjostrom, J. "Selecting Optimal Pipe Sizes For Water Distribution Systems". Journal, American Water Works Association, February 1988.

Whitworth, Robert. "Introduction to Manometers". Digester/Over the Spillway, September 1987.

INDEX